Exam

AS

Physics

Gurinder Chadha

Different types of questions

In AS examinations, there are mainly two sorts of examination questions: *structured* questions and *extended writing* questions.

Structured questions

In a structured question, a collection of common ideas is tested and the question is set out in smaller sub-sections. The earlier sections of the questions have a tendency to be easier and are designed to ease you into the question. Sometimes the answer to a particular section is given, so as to point you in the right direction. The later sections of a question can be a little discriminatory and you might have to think very carefully about your response. The following guide may be helpful:

- Read each question with great care and underline or highlight any key terms or data given.
- Write your answers carefully and always show **all** stages of your calculations.
- At every opportunity, keep referring to any introductory section to the question.

Extended questions

In an extended question, examiners will be assessing your understanding of physics and your ability to communicate ideas effectively and clearly. It is vital that in such questions, you pay particular attention to spellings, grammar and sentence construction. The response to such questions can be open-ended, with examiners keen to award marks for any 'good', but relevant, physics. The following guide may be helpful:

- Read the question carefully and make sure that you do clearly understand what the examiners want.
- Make a quick and short plan of your key ideas. It is worth putting your ideas in a *hierarchical* order.
- Write your answers clearly and remember to pay particular attention to the way you write.

Other questions

In addition to structured and extended questions, examiners also make use of free-response and open-ended questions. The points raised in the guide above are equally applicable here.

What examiners look for

An examination paper is devised to assess your knowledge, understanding and application of physics. One of the main purposes of an examination paper is to sort out how good you are at physics. Examiners are looking for the following points:

- Correct answers to the questions.
- Clearly presented answers with all the working shown.
- Concise answers to structured questions and logically set out responses to extended questions.
- Sketch graphs and diagrams that are drawn neatly, with particular attention given to labels and units.

What makes an A, C and E candidate?

The examiners can only ask questions that are within the specifications (this is the new term for the 'syllabus'). It is therefore vital that you are fully aware of what the examiners can and cannot ask in any examination. Be prepared to tackle the complexities of the paper. Your aim must be to achieve high marks in unit or module examinations. The way to accomplish this is to have a good knowledge and understanding of physics. Listed below are the minimum marks for grades A to E.

Grade A 80% **Grade B** 70% **Grade C** 60% **Grade D** 50% **Grade E** 40%

- **A grade candidates** have an excellent all-round knowledge of physics and they can apply that knowledge to new situations. Such candidates tend to be strong in all of the modules and tend to have excellent recall skills.
- **C grade candidates** have a reasonable knowledge of physics but have some problems when they apply their knowledge to new situations. They have some gaps in their knowledge and tend to be weak in some of the modules.
- **E grade candidates** have a poor knowledge of physics and have not learnt to apply their ideas to familiar and new situations. Such candidates find it difficult to recall key definitions and equations.

Revision skills

- Always start with a topic that you find easier. This will boost your self-confidence.
- Do not revise for too long. When you are tired and irritable, you cannot produce quality work.
- Make notes on post cards or lined paper of key ideas and equations. Do not feel that you have to write down everything. Just the key points need to be jotted down. Sometimes you have to learn certain proofs. It is worth writing down all the important steps for such proofs.
- Make good use of the specification. Use a highlighter pen to identify topics that you have already revised.
- Do not leave your revision to the last moment. Plan out a strategy spread over many weeks before the actual examination. Work hard during the day and learn to relax when needed.
- Whatever happens, do not try to learn any new topics on the day before the examination. It is important for you to be calm and relaxed for the actual examination.

Practice questions

This book is designed to improve your understanding of physics and, of course, improve your final grade.

Look carefully at the grade A and C candidates' responses. Can you do better? There are some important tips given to improve your understanding.

Try the practice examination questions and then look at the answers and tips given.

When you are ready, try the AS mock examination papers.

Planning and timing your answers in the exam

- Write legibly and stay focused throughout.
- Sometimes, candidates think that they have answered all the questions and then find an entire question on the last page. You do not want to be in this predicament, so **quickly** scan through the entire paper to see what you have to do.
- Do the question on the paper that you are most comfortable with. This will boost your confidence.
- Read each question carefully. Highlight the key ideas and data. The information given is there to be used.
- As a very rough guide, you have about 1 minute for each mark. The number of lines allocated for your answer gives you an idea of the depth and detail required for a particular answer. The marks allocated for each sub-section give an idea of how many steps or items of information are required.
- Do a quick plan for extended questions. It is not sensible to start writing straight away because you will end up either repeating yourself or missing out some important ideas.
- Do not use correction fluid. If you have strong reasons that a particular answer is wrong, then simply cross it out and provide an alternative answer.

Setting out numerical answers

It is important that your answers to numerical and algebraic questions are set out logically for the examiner. In this book, a simple method is used to indicate **where** a mark is awarded for the correct response. This is indicated by means of a tick ($\checkmark$).

The **marking scheme** adopted in this book and how you ought to **structure** a numerical answer is illustrated in the example below.

Question: What is the pressure exerted by a force of 9.0 kN acting on an area of $1.5 \times 10^{-2}\,\text{m}^2$? [2]

Answers:

$P = F/A$ $\checkmark$ (Make the physics clear to the examiner.)

$P = 9.0 \times 10^3 / 1.5 \times 10^{-2}$ (Use standard form and remember to convert k $\rightarrow 10^3$.)

$P = 6.0 \times 10^5\,\text{Pa}$ $\checkmark$ (Do not forget the correct unit and significant figures.)

There are only two marks for the calculation. One mark is awarded for the equation and the other for the correct answer and the unit.

Remember, the ticks appear next to responses where the marks are awarded.

No credit can be given for a **wrong** answer. However, by writing down all the stages of your work, it may be possible to pick up some or all of the 'part marks'. So help yourself and set out your work in a clear and methodical way.

What examiners look for

Examiners cannot give credit for the wrong physics. A wrongly quoted equation cannot be awarded any marks. For an examiner, it is quite disturbing to find candidates who cannot re-arrange equations. Mathematics is the language of physics, therefore it is important for candidates to be comfortable with handling and re-arranging equations. There are several techniques for re-arranging equations, but the one outlined below can be learnt and applied quickly.

Remember BODMAS.

When solving a numerical or algebraic equation, you must do the mathematical operations in the order given by the mnemonic BODMAS.

$$\text{Bracket} \rightarrow \text{Of} \rightarrow \text{Division} \rightarrow \text{Multiplication} \rightarrow \text{Addition} \rightarrow \text{Subtraction}$$

When it comes to re-arranging an equation, you simply **reverse** the mathematical operations. Here is an example to illustrate this technique for re-arranging an equation.

$v^2 = u^2 + 2as$ What is a?

Using the ideas developed above, we have

$$a \rightarrow (\times 2s) \rightarrow (+ u^2) \rightarrow = v^2$$

By reversing the sequence and carrying out the inverse operations, we end up with

$$v^2 \rightarrow (- u^2) \rightarrow (\div 2s) \rightarrow = a$$

Therefore $a = \dfrac{(v^2 - u^2)}{2s}$

If you have some other tried and tested technique for re-arranging equations, then it is best to stick to it. However, do remember to take re-arranging of equations seriously in physics.

Here are some other suggestions to **boost your grade**.

- Be familiar with the specifications.

- Learn all the definitions within the specifications. Recalling definitions can give you easy marks and improve your final grade.

- Write all the stages of a numerical solution. If your final answer is wrong, you still have a chance to pick up some of the 'part-marks'.

- In a question with 'state', the answer is brief and does not require any further explanation. In a question with 'describe', the answer can be long and may require full explanation of some physics.

- Use the information given in the question to guide your answers. For numerical solutions, keep an eye on the significant figures and units. Your final answer must not be more or less the significant figures given in the question. In physics, it is sensible to write the final answer in standard form, e.g. 1.62×10^{-3} A.

- Read the information given on graphs and tables carefully. Sometimes data is given in either standard or prefix form. Do not forget to take this on board when doing your calculations. For example, the stress axis is labelled as 'stress/MPa'. Remember that 'MPa' is 10^6 Pa.

- It is easy to press the wrong buttons on the calculator. Make sure that your answer looks reasonable. If you have time, you ought to check a calculation again.

- Draw diagrams carefully and make sure that you label all the key items.

- Your graphs must be correctly labelled and have a suitable scale so that they fill most of the graph paper.

- You do not have to recall physical data. All the data required is normally given on the question paper itself or on a separate data-sheet.

For more information about your course go to:
www.aqa.org.uk   www.ccea.org.uk
www.edexcel.org.uk   www.wjec.org.uk
www.ocr.org.uk

Questions with model answers

C grade candidate – mark scored 6/10

Examiner's Commentary

1 Explain what is meant by an **elastic** material. [1]

A material that returns to its original shape and size when the forces are removed. ✔

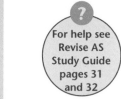

For help see Revise AS Study Guide pages 31 and 32

2 With the aid of a sketch graph, show the force against extension graph for a typical ductile material. [2]

[sketch graph: force (vertical axis) against extension (horizontal axis), showing a straight line from origin] ✔

There is no evidence of what happens to the material beyond the elastic limit, hence the candidate lost a valuable 'detail' mark. If the candidate had shown plastic deformation, then the second mark could have been scored.

3 One of many cables supporting a small suspension bridge has a radius of 1.5 cm and a natural length of 9.0 m. The tension in each cable is 65 kN. The Young modulus of the cable material is 2.1×10^{11} Pa.

(a) Show that the cable has a cross-sectional area of $7.1 \times 10^{-4}\,m^2$. [1]

$$\text{area} = \pi r^2$$
$$\text{area} = \pi \times (1.5)^2 = 7.07 \ ✗$$

The candidate did not convert the radius into metres, the answer is therefore out by a factor of 10^4.

(b) Calculate the stress in each cable. [2]

$$\text{stress} = \frac{F}{A}$$
$$\text{stress} = \frac{65 \times 10^3}{7.1 \times 10^{-4}} \ ✔$$
$$\text{stress} = 9.2 \times 10^7 \, Pa \ ✔$$

Fortunately, the candidate used the information given in **(a)** and that was a sensible strategy.

(c) Calculate the strain experienced by the cable and its extension. State any assumption made. [4]

$$E = \frac{\text{stress}}{\text{strain}}$$
$$\text{strain} = \frac{\text{stress}}{E}$$
$$\text{strain} = \frac{9.2 \times 10^7}{2.1 \times 10^{11}}$$
$$\text{strain} = 4.4 \times 10^{-4} \ ✔$$

The material obeys Hooke's law. ✔

The question also required the extension of the cable, this was not done by the candidate. Since
$$\text{strain} = \frac{\text{extension}}{\text{original length}}$$
the extension can be calculated from the value of the strain. The correct value for the extension is $3.9 \times 10^{-3}\,m$.

GRADE BOOSTER

This candidate could have improved the grade by using SI units and reading question **3(c)** more carefully.

Examiner's Commentary

The diagram shows a person pushing a roller at a constant velocity of 0.9 m s^{-1} along a flat horizontal ground. The handle makes an angle of 37° to the horizontal and a force of 60 N is directed along the handle as shown on the diagram.

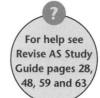

For help see Revise AS Study Guide pages 28, 48, 59 and 63

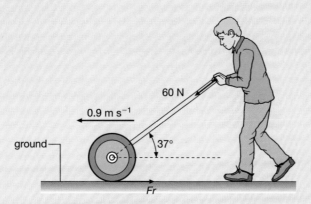

60 N

0.9 m s^{-1}

ground

37°

Fr

1 Calculate the horizontal component of the force acting on the roller. [2]

$F_x = F \cos\theta$ ✔

$F_x = 60 \times \cos 37 = 48$ N ✔

2 What is the magnitude of the frictional force F_r on the roller? Explain your answer. [2]

Friction F_r is equal to 48 N ✔

because there is no force on the roller. ✗

The roller is moving at a constant velocity and therefore has no acceleration. According to $F = ma$, there is no **net** force acting on the roller. The candidate's answer is wrong because there are several forces acting on the roller. It so happens that horizontally, there is no **resultant** force.

3 Calculate the rate of work done in pushing the roller along the ground. [3]

work = Fx

In one second, the distance moved is 0.9 m.

Work done in 1s is the power. Therefore

power = 48 × 0.9 ✔

power = 43 W ✔ ✔

One mark was reserved for the correct unit for the rate of working or the power.

Force, motion and energy

1 (a) Distinguish between a **scalar** quantity and a **vector** quantity. [2]

(b) A car travels one complete lap around a circular track of radius 500 m in a time of 2.0 minutes. Calculate

    (i) its average speed, [2]

    (ii) its displacement from the starting point after 0.5 minutes. [2]

[Total: 6]

---

2 An aircraft of total mass 1.5×10^5 kg accelerates, at maximum thrust from the engines, from rest along a runway for 25 s before reaching the required speed for take-off of 65 m s^{-1}. Assume the acceleration of the aircraft is constant.
Calculate

(a) the acceleration of the aircraft, [3]

(b) the force acting on the aircraft to produce this acceleration, [2]

(c) the distance travelled by the aircraft in this time. [2]

[OCR June 2003]

[Total: 7]

Answers on pages 16–21 **Answers** on pages 16–21 **Answers** on pages 16–21

3 **(a)** State the principle of moments. [2]

(b) The diagram shows a **model** for a human arm that is balanced in the horizontal position.

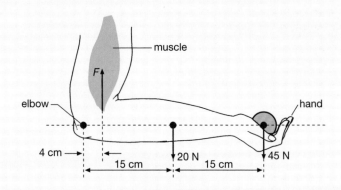

The arm is assumed to be uniform and its centre-of-gravity is 15 cm away from the elbow. The weight of the arm is 20 N and the hand is holding an object of weight 45 N.

  (i) Explain what is meant by *centre-of-gravity*. [1]

 (ii) Calculate the vertical force *F* provided by the arm muscle. [3]

(iii) The arm is extended further away from the body but still balanced in the horizontal position. The arm muscle now exerts a force *F* at an angle to the vertical. This is shown in the diagram below.

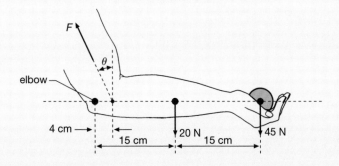

Without any further calculations, explain how the magnitude of the force *F* changes as compared with your answer to **(b)(ii)**. [2]

[Total: 8]

Answers on pages 16–21 **Answers** on pages 16–21 **Answers** on pages 16–21

4 The diagram below shows a lamp of total mass 320 g hung from the ceiling and supported by a horizontal plastic cord that is attached to a wall.

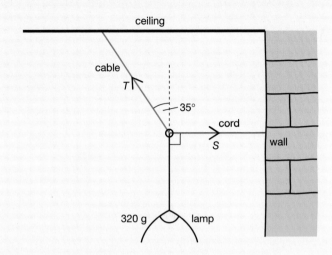

The lamp is in static equilibrium. The cable shown on the diagram makes an angle of 35° with the vertical.

(a) Explain what is meant by *static equilibrium*. [1]

(b) Show that the tension T in the cable is 3.8 N.
Data: $g = 9.8\,\text{N kg}^{-1}$ [2]

(c) The plastic cord has a diameter of 1.8 mm. Calculate

　　(i) the tension S in the cord, [3]

　　(ii) the stress experienced by the cord. [3]

[Total: 9]

5 (a) State Hooke's law. [2]

(b) The area under a force against extension graph is equal to work done by the force. Use this idea and the sketch graph below to find an expression for the energy stored in a spring in terms of the applied force F and the final extension e.

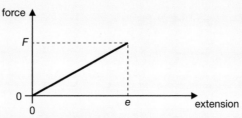

[2]

(c) A diagram of a child's toy is shown below.

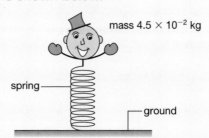

The total mass of the toy and spring is 4.5×10^{-2} kg. A force of 12 N compresses the spring by 5.0 cm.

(i) Calculate the energy stored in the spring when it is compressed by 5.0 cm. [2]

(ii) When the spring is released, the toy lifts off the ground and reaches a maximum vertical height h above the ground. Calculate this height h.
State any assumption made.
Data: $g = 9.8$ N kg^{-1} [3]

[Total: 9]

6 (a) Define acceleration. [1]

(b) A 1200 kg car is travelling towards a safety barrier at a velocity of 28 m s^{-1}. The diagram shows the velocity against time graph for the car when colliding with this barrier.

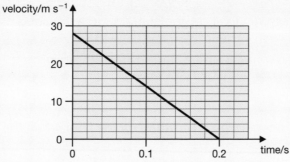

(i) Describe the motion of the car as it hits the safety barrier. [2]

(ii) Calculate the magnitude of the force exerted by the safety barrier on the car. [3]

(iii) Safety barriers and cars are designed to crumple on impact.
What effect does this have on car safety? [2]

[Total: 8]

Answers on pages 16–21 Answers on pages 16–21 Answers on pages 16–21

Force, motion and energy

7 In an accident on a motorway, a car of mass 950 kg leaves a skid mark 20 m long when stopping. The accident investigators suspect the deceleration of the car to be $12\,\mathrm{m\,s^{-2}}$.

(a) Calculate the magnitude of the average braking force between each of the four car tyres and the road. [2]

(b) Calculate the initial speed of the car. [2]

[Total: 4]

8 **(a)** Write an equation for the kinetic energy of an object. Define any symbols used. [2]

(b) A 3.2×10^{-2} kg metal ball is dropped from a vertical height 2.0 m. It hits the ground and makes a dent of depth 4.0 mm.

(i) Show that the impact velocity of the metal ball with the ground is $6.3\,\mathrm{m\,s^{-1}}$. You may assume that there is negligible air resistance.
Data: $g = 9.8\,\mathrm{m\,s^{-2}}$ [2]

(ii) Calculate the kinetic energy of the metal ball just before it hits the ground. [2]

(iii) Calculate the mean deceleration of the ball during its impact with the ground. [2]

[Total: 8]

9 The diagram shows an 8.0 kg shopping bag placed on the floor of a lift.

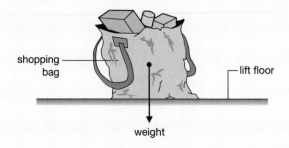

(a) Complete the diagram to show the contact force provided by the floor on the bag. Label the force R. [1]

(b) Calculate the weight of the shopping bag.
Data: $g = 9.8\,\mathrm{N\,kg^{-1}}$ [1]

(c) Calculate the contact force R provided by the floor when the lift is moving vertically upwards

(i) at a constant velocity of $0.50\,\mathrm{m\,s^{-1}}$, [1]

(ii) at an acceleration of $2.5\,\mathrm{m\,s^{-2}}$. [2]

[Total: 5]

Answers on pages 16–21 **Answers** on pages 16–21 **Answers** on pages 16–21

10 The diagram shows an aircraft travelling horizontally at a constant velocity of 65 m s^{-1} at a height of 80 m above the ground.

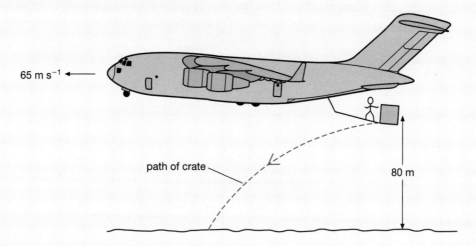

A 620 kg crate is dropped from the aircraft from this height.
Data: $g = 9.8\,\text{N}\,\text{kg}^{-1}$

(a) Calculate the gravitational potential energy of the crate before it is dropped from the aircraft. [3]

(b) Calculate the time taken for the crate to reach the ground. (You may ignore the effects of air resistance on the motion of the crate.) [3]

(c) Calculate the magnitude and the direction of the velocity of the crate when it hits the ground. [4]

[Total: 10]

11 The diagram shows a car hanging from a helicopter.

cable

1200 kg

(a) The helicopter is hovering at a constant height above the ground. The mass of the car is 1200 kg.

    (i) Calculate the tension in the cable supporting the car.
       Data: $g = 9.8\,N\,kg^{-1}$ **[2]**

    (ii) The cable is made from a material of Young modulus 1.9×10^{11} Pa and has a cross-sectional area of $5.0 \times 10^{-5}\,m^2$. The original length of the cable was 14 m. Calculate the extension of the cable assuming that it has not exceeded its elastic limit.   **[4]**

(b) When the helicopter is flying horizontally, the cable supporting the car makes an angle with the vertical. State and explain the effect on your answer to **(a)(i)**.   **[2]**

                                         [Total: 8]

12 The diagram shows a stunt person of mass 80 kg skidding down a small ramp.

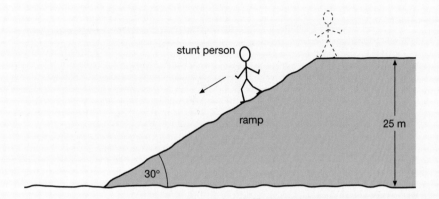

The person is at rest at the top of the ramp. At the bottom of the ramp, the speed of the person is 14 m s^{-1}.

(a) Calculate the kinetic energy of the person at the bottom of the ramp. [2]

(b) Show that, as the person skidded from the top to the bottom of the ramp, the work done against friction is 1.2 × 10^4 J.
Data: $g = 9.8$ N kg^{-1} [3]

(c) Use your answer to **(b)** to determine the magnitude of the resistive force acting on the person whilst skidding down the ramp. (You may assume that this force is constant.) [3]

[Total: 8]

(1) (a) A scalar quantity has only size (or magnitude).
A vector quantity has both magnitude **and** direction.

(b) **(i)** speed = distance / time
$$\text{speed} = \frac{2\pi \times 500}{2.0 \times 60}$$
speed = $26.2\,\mathrm{m\,s^{-1}} \approx 26\,\mathrm{m\,s^{-1}}$

(ii) displacement = $\sqrt{500^2 + 500^2} \approx 710\,\mathrm{m}$
The displacement is at an angle of 45° from the initial direction of travel.

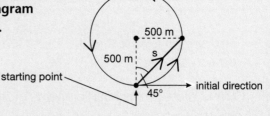
(2) (a) $a = \dfrac{v - u}{t} = \dfrac{65 - 0}{25}$
$a = 2.6\,\mathrm{m\,s^{-2}}$

(b) $F = ma = 1.5 \times 10^5 \times 2.6$
$F = 3.9 \times 10^5\,\mathrm{N}$

(c) $s = \frac{1}{2}(u + v)t = \frac{1}{2}(0 + 65) \times 25$
$s = 813\,\mathrm{m} \approx 810\,\mathrm{m}$

(3) (a) For **rotational** equilibrium,
sum of the clockwise moments = sum of anticlockwise moments.
The moments are taken about the same point or pivot.

(b) **(i)** The centre-of-gravity is the point at which the weight of the object appears to act.

(ii) Take moments about the elbow.
sum of anticlockwise moments = sum of clockwise moments
$4 \times F = (15 \times 20) + (45 \times 30)$
$F = \dfrac{1650}{4} = 413 \approx 410\,\mathrm{N}$

(iii) The force *F* is greater than before.
The anticlockwise moment must still be the same. Hence, the **vertical** component of the force (given by $F\cos\theta$) must be 410 N as in **(b)(ii)**. This implies that the force *F* exerted by the muscle must be greater than before.

(4) (a) The **net** force acting on the lamp is zero. It therefore has no acceleration and remains at rest.

(b) Net force in the vertical direction is zero.

$T \cos 35° = $ weight of lamp

$T \cos 35° = 0.320 \times 9.8$

$T = \dfrac{0.320 \times 9.8}{\cos 35} = 3.83\,\text{N} \approx 3.8\,\text{N}$

(c) (i) Net force in the horizontal direction is zero.

$S = T \sin 35°$

$S = 3.83 \times \sin 35° = 2.197\,\text{N} \approx 2.2\,\text{N}$

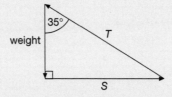
(ii) $\text{stress} = \dfrac{\text{force}}{\text{area}}$

$\text{stress} = \dfrac{2.197}{\pi \times 0.0009^2}$

$\text{stress} \approx 8.6 \times 10^5\,\text{Pa}$

(5) (a) The extension of a material is directly proportional to the applied force. This is true as long as the elastic limit (or limit of proportionality) is not exceeded.

(b) $\text{Area} = \frac{1}{2} \times \text{base} \times \text{height}$ (area of a triangle)

$\text{energy} = \frac{1}{2} Fe$

(c) (i) $\text{energy} = \frac{1}{2} Fe$

$\text{energy} = \frac{1}{2} \times 12 \times 5.0 \times 10^{-2}$

$\text{energy} = 0.30\,\text{J}$

(ii) Assumption: gravitational P.E. = energy stored in spring

$mg\Delta h = 0.30$

$h = \dfrac{0.30}{4.5 \times 10^{-2} \times 9.8}$

$h \approx 0.68\,\text{m}$

Force, motion and energy

(6) (a) acceleration = rate of change of velocity

(b) **(i)** The car experiences a **constant** deceleration.

> **examiner's tip**
>
> The gradient from a velocity–time graph is equal to acceleration. The gradient of the line is constant but negative. Therefore, the car's velocity is decreasing at a constant rate.

(ii) acceleration = gradient

$$a = -\frac{28}{0.2} = -140 \text{ m s}^{-2}$$

$F = ma$

$F = 1200 \times 140 = 1.68 \times 10^5 \approx 1.7 \times 10^5 \text{ N}$ \qquad (magnitude only)

> **examiner's tip**
>
> The magnitude of the force is required, so the negative sign in the answer is optional.

(iii) Impact forces are smaller because the time of impact is longer (hence smaller deceleration).

---

(7) (a) $F = ma$

$F = 950 \times 12$ \qquad (magnitude only)

$F = 1.14 \times 10^4 \text{ N}$

Force on each tyre $= \dfrac{1.14 \times 10^4}{4}$

Force on each tyre $= 2.85 \times 10^3 \approx 2.9 \times 10^3 \text{ N}$

(b) $v^2 = u^2 + 2as$

$a = -12 \text{ m s}^{-2}$ \qquad $v = 0$ \qquad $s = 20 \text{ m}$ \qquad $u =$ initial velocity

$u^2 = 2 \times 12 \times 20$

$u = 21.9 \approx 22 \text{ m s}^{-1}$

> **examiner's tip**
>
> Instead of kinematics, you can solve the problem using energy considerations.
> Work done to stop car = Initial K.E. of car
>
> $Fs = \frac{1}{2}m u^2$
>
> $u = \sqrt{\dfrac{2Fs}{m}}$
>
> $u = \sqrt{\dfrac{2 \times 1.14 \times 10^4 \times 20}{950}}$
>
> $u = 22 \text{ m s}^{-1}$

Force, motion and energy

(8) (a) K.E. $= \frac{1}{2}mv^2$

m is the mass of object and v is its speed.

(b) (i) $v^2 = u^2 + 2as$

$u = 0$, therefore $v^2 = 2as$

$v = \sqrt{2as} = \sqrt{2 \times 9.8 \times 2.0}$

$v = 6.26\,\text{m s}^{-1}$

(ii) K.E. $= \frac{1}{2}mv^2$

K.E. $= \frac{1}{2} \times 3.2 \times 10^{-2} \times 6.26^2$

K.E. $= 0.627 \approx 0.63\,\text{J}$

(iii) Work done = K.E. of the ball

$Fs = 0.627$

(F = force on the ball)

$F = \dfrac{0.627}{4.0 \times 10^{-3}}$

$F = 1.57 \times 10^2 \approx 1.6 \times 10^2\,\text{N}$

$a = \dfrac{F}{m}$ (magnitude only)

$a = \dfrac{1.57 \times 10^2}{3.2 \times 10^{-2}}$

$a = 4.91 \times 10^3 \approx 4.9 \times 10^3\,\text{m s}^{-2}$

Force, motion and energy

(9) (a) The contact force R is vertically upwards.

(b) weight $= mg$
weight $= 8.0 \times 9.8 = 78.4 \approx 78$ N

(c) (i) $F = ma$
Since $a = 0$, the **net** force F on the bag $= 0$
Therefore $R =$ weight
$R \approx 78$ N

(ii) Net force vertically $= ma$
$R - W = 8.0 \times 2.5$ ($W =$ weight)
$R - 78.4 = 20$
$R = 98.4 \approx 98$ N

examiner's tip

It is very tempting to just use '$R = ma$'. This is, of course, not the case. The force R must be greater than the downward force W in order to provide the bag with an upward acceleration.

(10) (a) P.E. $= mgh$
P.E. $= 620 \times 9.8 \times 80$
P.E. $= 4.86 \times 10^5$ J $\approx 4.9 \times 10^5$ J

(b) $s = ut + \frac{1}{2}at^2$ ($u = 0$)
$80 = \frac{1}{2} \times 9.8 \times t^2$
$t = \sqrt{\dfrac{2 \times 80}{9.8}} = 4.04$ s ≈ 4.0 s

examiner's tip

This is a question on 'projectiles'. The vertical component of the velocity is affected by the Earth's gravitational field but the horizontal component of the velocity remains constant.

(c) Horizontal component of the velocity, $v_x = 65$ m s^{-1}
Vertical component of the velocity, $v_y = 0 + 9.8 \times 4.04 = 39.6$ m s^{-1}

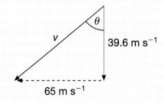

Magnitude of impact velocity, $v = \sqrt{(65^2 + 39.6^2)} = 76$ m s^{-1}

$\tan \theta = \dfrac{65}{39.6} \Rightarrow \theta = 59°$

examiner's tip

Do not forget that velocity is a vector. The question therefore wants you to work out the magnitude **and** the direction of the velocity.

(11) (a) (i) tension = weight of car

tension = mg = 1200 × 9.8

tension = $1.176 \times 10^4 \approx 1.18 \times 10^4$ N

(ii) stress = $\dfrac{\text{force}}{\text{area}}$

stress = $\dfrac{1.176 \times 10^4}{5.0 \times 10^{-5}} = 2.35 \times 10^8$ Pa

$E = \dfrac{\text{stress}}{\text{strain}}$

strain = $\dfrac{2.35 \times 10^8}{1.9 \times 10^{11}} = 1.24 \times 10^{-3}$

strain = $\dfrac{\text{extension}}{\text{original length}}$

extension = $1.24 \times 10^{-3} \times 14$

extension = 1.74×10^{-2} m ≈ 1.7 cm

(b) The upward vertical force will be equal to the vertical component of the tension T in the cable ($T\cos\theta$). This component must be equal to the weight W of the car.

$T\cos\theta = W$

$T = \dfrac{W}{\cos\theta} > W$

The cosine of an angle is always less than 1, hence the tension is greater than the weight of the car.

---

(12) (a) K.E. $= \frac{1}{2}mv^2$

K.E. $= \frac{1}{2} \times 80 \times 14^2$

K.E. $= 7.84 \times 10^3$ J $\approx 7.8 \times 10^3$ J

(b) The work done against friction is equal to the **difference** between the gravitational P.E. at top of ramp and the K.E. at bottom of ramp.

work done against friction = W

$W = mgh - 7.84 \times 10^3$

$W = (80 \times 9.8 \times 25) - 7.84 \times 10^3$

$W = 1.18 \times 10^4$ J $\approx 1.2 \times 10^4$ J

(c) work done = force × distance

distance travelled against the friction $= \dfrac{25}{\sin 30} = 50$ m

force $= \dfrac{1.18 \times 10^4}{50} \approx 240$ N

examiner's tip	Do not forget that work done by (or against) a force requires the distance travelled in the **direction** of the force. To determine the magnitude of the resistive force you do need the correct distance. You would lose some marks for using the 25 m given on the diagram.

Questions with model answers

C grade candidate – mark scored 6/10

Examiner's Commentary

1 Complete the sentence below.

Electric current is equal to the rate of flow of charge ✔ [1]

> The answer is correct. A shorter but alternative route would be to use
>
> $$P = \frac{V^2}{R}$$
>
> $$R = \frac{V^2}{P}$$
>
> $$\therefore R = \frac{240^2}{1000}$$
>
> $$R = 58\,\Omega$$

2 Show that the alternative unit for current is $C\,s^{-1}$. [1]

$$I = \frac{\Delta Q}{\Delta t} \qquad I \rightarrow [coulombs\ /\ seconds]\ ✔$$

Hence, current $\rightarrow [C\ s^{-1}]$

3 A washing machine, rated as '240 V, 1 kW', is operated for 30 minutes.

(a) Calculate the resistance of the appliance. [2]

$$P = VI \qquad I = \frac{P}{V} = \frac{1000}{240} \quad \therefore I = 4.17\ A\ ✔$$

$$V = IR \qquad\qquad (Ohm's\ law)$$

$$R = \frac{V}{I} = \frac{240}{4.17} \qquad\qquad R = 58\,\Omega\ ✔$$

> The candidate managed to secure one mark for calculating the charge flow in the time of 30 minutes. It was sensible to convert the time into seconds. The total charge of 7.5×10^3 C is due to N number of electrons, each carrying a charge of 1.6×10^{-19} C. Therefore:
>
> $$7.5 \times 10^3 = N\,e$$
>
> $$N = \frac{7.5 \times 10^3}{1.6 \times 10^{-19}}$$
>
> $$N = 4.7 \times 10^{22}$$

(b) Calculate the charge flow through the appliance in a period of 30 minutes. Hence determine the number of electrons responsible for this flow of charge.

Data: $e = 1.6 \times 10^{-19}\,C$ [3]

$$\Delta Q = I\,\Delta t$$

$$\Delta Q = 4.17 \times (30 \times 60) = 7.5 \times 10^3\ C\ ✔$$

number of electrons $= 7.5 \times 10^3 \times 1.6 \times 10^{-19}$ ✗

number $= 1.2 \times 10^{-15}$ ✗

(c) What is the cost of using the appliance for 30 minutes? The cost of each kW h of energy is 6.4p. [3]

$$E = P\Delta t \qquad E = 1000 \times (30 \times 60) = 1.8 \times 10^6\ J\ ✔$$

$$1\ kW\ h = 1 \times 3600 = 3600\ J\ ✗$$

$$\therefore cost = \frac{3600 \times 6.4}{1.8 \times 10^6}$$

$$= 0.013\ p\ ✗$$

> The amount of energy transformed by the appliance is correct. The number of 'units' of energy transformed is given by
>
> number of 'units' = 1 kW × (30/60) h
>
> = 0.50 kW h
>
> ∴ cost = 0.50 × 6.4 = 3.2 p

? For help see Revise AS Study Guide pages 74 and 76

GRADE BOOSTER

You cannot afford to have any 'gaps' in your knowledge of AS Physics. Learn all the material on the syllabus/specification. This candidate lost valuable marks by not having detailed knowledge of the kilowatt hour.

The diagram shows an electrical circuit based on two switches A and B. The supply may be assumed to have negligible internal resistance.

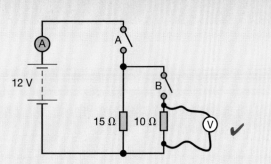

Examiner's Commentary

For help see Revise AS Study Guide pages 71, 76 and 77

1 On the diagram, show how the potential difference across the 10 Ω resistor may be measured. [1]

> With the switch B open, there can be no current in the 10 Ω resistor. Before substituting any numbers into the $V = IR$ equation, it is worth spending a few moments scrutinising the circuit.

2 Calculate the ammeter reading when A is **closed** and B is **open**. [2]

Current is only in the 15 Ω resistor.

$$I = \frac{V}{R} = \frac{12}{15} \quad ✔$$

$$I = 0.8 \text{ A} \quad ✔$$

3 A and B are **both** closed.

> Under the time restrictions of an exam, candidates often forget to inverse their answer when using
> $$\frac{1}{R} = \frac{1}{R_1} + \frac{1}{R_2}.$$
> To avoid this, another equivalent equation for determining the total resistance may be used. For **two** resistors of resistance values R_1 and R_2 connected in parallel, the total resistance R_T, is given by
> $$R_T = \frac{R_1 R_2}{(R_1 + R_2)}$$
> Therefore:
> $$R_T = \frac{15 \times 10}{(15 + 10)}$$
> $$R_T = 6.0 \, \Omega$$

(a) Calculate the new ammeter reading. [3]

$$\frac{1}{R} = \frac{1}{15} + \frac{1}{10} \quad ✔$$

$$\frac{1}{R} = \frac{25}{150} = 0.167$$

$$R = \frac{1}{0.167} = 6.0 \, \Omega \quad ✔$$

$$I = \frac{V}{R} = \frac{12}{6.0}$$

$$I = 2.0 \text{ A} \quad ✔$$

(b) Calculate the ratio

$$\frac{\text{power dissipated in 15 Ω resistor}}{\text{power dissipated in 10 Ω resistor}}$$ [2]

$$P = \frac{V^2}{R}$$

For 10 Ω resistor: $\quad P = \frac{12^2}{10} = 14.4 \text{ W}$

For 15 Ω resistor: $\quad P = \frac{12^2}{15} = 9.6 \text{ W} \quad ✔$

$$\therefore \text{Ratio} = \frac{14.4}{9.6} = 1.5 \quad ✗$$

> After all the hard work, the candidate calculated the ratio incorrectly. When calculating the electrical power, the candidate has used 12 V for both resistors. Since the p.d. across a parallel circuit is the same, it follows that $P \propto \frac{1}{R}$.
> The ratio will therefore be:
> $$\text{ratio} = \frac{10}{15}$$
> $$\text{ratio} = 0.67$$

Electricity

1 **(a)** Explain what is meant by electric current. [1]

(b) With the aid of a diagram, show how you would determine the electrical resistance of a resistor using meters and a battery. [3]

(c) **(i)** State Ohm's law. [2]

(ii) Name a component that does **not** obey Ohm's law. [1]

(iii) 1. Sketch the current–voltage characteristic of a semiconducting diode.

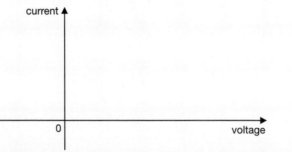

[2]

2. Explain how the resistance of the semiconducting diode changes as the voltage (potential difference) is increased from negative to positive values. [3]

[Total: 12]

2 A sheet of carbon-reinforced plastic measuring 80 mm × 80 mm × 1.5 mm has its two large surfaces coated with highly conducting metal film. When a potential difference of 240 V is applied between the metal films, there is a current of 2.0 mA in the plastic. Calculate the resistivity of the plastic. [3]

[AQA June 2002]

[Total: 3]

3 **(a)** The current I in a metallic wire is given by the equation

$$I = Anev$$

Define the symbols used in this equation. [2]

(b) A metallic material is connected to a battery. The cross-sectional shape of the material is shown below.

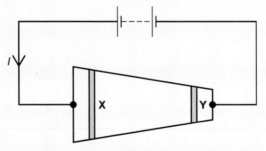

(i) Suggest why the current in sections **X** and **Y** is the **same**. [1]

(ii) State and explain what happens to the mean drift velocity of the electrons at **Y** compared with that at **X**. [2]

(iii) Explain why the section **Y** is warmer than the section **X**. [2]

[Total: 7]

Answers on pages 28–33 **Answers** on pages 28–33 **Answers** on pages 28–33

4 The diagram shows a circuit for a simple electrical thermometer.

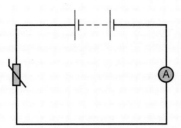

The ammeter, connected in series with the negative temperature coefficient (NTC) thermistor, has negligible resistance. Explain the effect on the ammeter reading as the temperature of the thermistor is increased. [3]

[Total: 3]

---

5 The maximum power a 120 Ω resistor can dissipate is 0.50 W.
For this resistor, calculate

(a) the maximum current in it, [2]

(b) the maximum potential difference that can be safely applied. [2]

[Total: 4]

---

6 **(a)** Define electrical resistance. [1]

(b) Show that electrical resistivity has the unit Ω m. [2]

(c) A car lamp is rated as '12 V, 60 W'. The lamp has a filament wire of radius 3.2×10^{-4} m and when operating at 12 V, the metal of the filament has resistivity 4.2×10^{-7} Ω m.

   (i) Calculate the resistance of the lamp when operated at 12 V. [2]

   (ii) Calculate the length of the filament wire and comment on its value. [3]

[Total: 8]

Answers on pages 28–33 Answers on pages 28–33 Answers on pages 28–33

7 The diagram shows a potential divider circuit based on a light-dependent resistor (LDR). Assume the supply has negligible internal resistance.

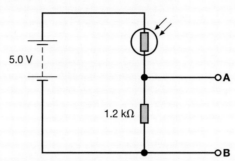

(a) State how the resistance of the LDR is affected by the intensity of light falling on it. **[1]**

(b) At a particular intensity of light, the LDR has a resistance of 820 Ω.
Calculate the potential difference between **A** and **B**. **[3]**

(c) On the axes provided, draw a sketch graph to show the variation of the potential difference V between **A** and **B** with light intensity.

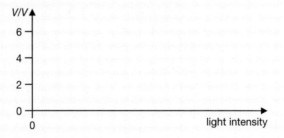

[2]

[Total: 6]

---

8 The diagram shows a cell of e.m.f. 1.2 V and internal resistance 0.8 Ω is connected in a circuit with three resistors, each having a resistance of 4.7 Ω.

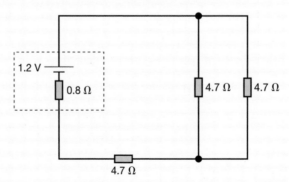

Calculate

(a) the current drawn from the cell, **[3]**

(b) the potential difference across the terminals of the battery, **[2]**

(c) the fraction of the total power lost within the battery due to its internal resistance. **[3]**

[Total: 8]

Answers on pages 28–33 Answers on pages 28–33 Answers on pages 28–33

9 The diagram shows a circuit designed to locate the position of a metal rod that rolls along two parallel resistance wires.

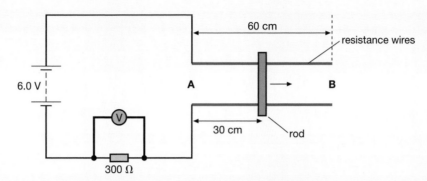

The battery may be assumed to have negligible internal resistance and the voltmeter has an infinite resistance. Each resistance wire has a total length of 60 cm and a resistance per unit length of 15 $\Omega\,\text{cm}^{-1}$. The metal rod has negligible resistance.

(a) Calculate the total resistance of each wire. [1]

(b) Calculate the voltmeter reading when the metal rod is at the position shown in the diagram. [3]

(c) State and explain how the voltmeter reading would change as the metal rod rolls from **A** towards **B**. [2]

[Total: 6]

10 **(a)** State Kirchhoff's first and second laws. [2]

(b) The diagram shows an electrical circuit.

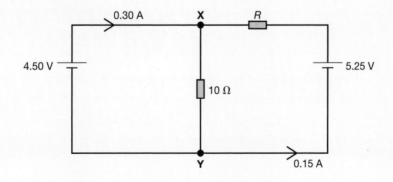

Each battery has negligible internal resistance.

   (i) State the current in the 10 Ω resistor. [1]

   (ii) Determine the resistance R of the resistor. [3]

[Total: 6]

Answers on pages 28–33 **Answers** on pages 28–33 **Answers** on pages 28–33

Electricity

(1) (a) Electric current is the flow of charge.

(b) The circuit is shown below.

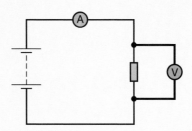

The voltmeter reading gives the p.d. across the resistor and the current recorded by the ammeter gives the current in the resistor. The resistance of the resistor is determined using the equation:

$$\text{resistance} = \frac{\text{potential difference}}{\text{current}}$$

examiner's tip

It is best to use accepted electrical symbols for the meters. The voltmeter may also be connected across the battery if it were to have negligible internal resistance. Since the question indicates nothing about the internal resistance of the battery or the ammeter, it is best to place the voltmeter across the component itself.

(c) (i) The current is directly proportional to the potential difference for a metallic conductor at constant temperature.

examiner's tip

There is one mark reserved here for 'detail'. Many candidates tend to write
current ∝ voltage
but do not qualify that this is only true for metals kept at the same physical conditions. Look carefully at the number of marks reserved for the answer. Two marks would automatically imply two distinct items for credit.

(ii) A filament lamp or a semiconducting diode.

(iii) 1.

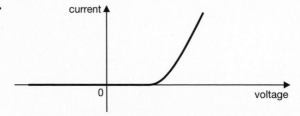

No current shown for negative values of voltage.
An increasing current shown for positive values of voltage.

examiner's tip

A semiconducting diode does not conduct immediately when voltage > zero. There is a 'switch on' voltage depending on the semiconducting material. A silicon diode will start to conduct when the p.d. across it is about 0.6 V. A germanium diode on the other hand, starts to conduct when the p.d. is about 0.2 V. Provide as much detail as possible in your answer.

2. For negative voltages, there is no current because the diode has infinite resistance.

For positive voltages, the resistance of the diode decreases as the voltage across it increases.

For a silicon diode, the resistance is infinite for voltage less than about $0.6\,V$.

---

(2) $R = \dfrac{V}{I} = \dfrac{240}{2.0 \times 10^{-3}} = 1.2 \times 10^5\,\Omega$

$\rho = \dfrac{RA}{L} = \dfrac{1.2 \times 10^5 \times 80 \times 80 \times 10^{-6}}{1.5 \times 10^{-3}} \approx 5.1 \times 10^5\,\Omega\,m$

examiner's tip

Always read the question carefully. The 'length' of the plastic sample is 1.5 mm and not 80 mm.

(3) (a) $A \rightarrow$ **cross-sectional** area of the wire.

$n \rightarrow$ number density of charges or number of charge carriers per unit volume.

$e \rightarrow$ elementary or electronic charge.

$v \rightarrow$ mean drift velocity of the charge carriers.

examiner's tip

There are four labels to be defined and only two available marks. Therefore, if there are two or more wrong answers, then most likely, the candidate will be given no mark.

(b) (i) The current is the same because both sections **X** and **Y** are connected in **series**.

(ii) The mean drift velocity at **Y** is greater than that at **X**.

$$I = Anev$$

Since I, n and e are the same for the material, we have

$$v \propto \frac{1}{A}.$$

The section **Y** has a smaller cross-sectional area and therefore the mean drift velocity at **Y** is greater.

examiner's tip

It is tempting to just use words to describe what is happening. It is, however, easier to convey the physics by using the equation

$$I = Anev$$

(iii) The section **Y** is warmer than **X** because its resistance (per unit length) is larger.

Power dissipated is given by $P = I^2 R$.

Since current I is the same, $P \propto R$.

Therefore section **Y** is warmer than **X**.

examiner's tip

This is a discriminating question asking the candidate to search through several ideas in physics. It requires a knowledge of $R = \dfrac{\rho \ell}{A}$, $P = I^2R$ and series circuits. The resistance of section Y is greater than that of X because

$$R \propto \frac{1}{A}.$$

(4) Increasing the temperature will **decrease** the resistance of the thermistor.

The current I in the thermistor is given by

$$I = \frac{V}{R}.$$

The p.d. across the thermistor remains constant. Hence the current in the thermistor will increase. (Note, in this case, $I \propto \frac{1}{R}$.)

---

(5) (a) $P = I^2 R$

$$0.50 = I^2 \times 120$$

$$I = \sqrt{\frac{0.50}{120}} = 6.455 \times 10^{-2}\,\text{A} \approx 65\,\text{mA}$$

(b) $P = VI$

$$V = \frac{P}{I} = \frac{0.50}{6.455 \times 10^{-2}}$$

$$V \approx 7.7\,\text{V}$$

examiner's tip You can also use $P = \dfrac{V^2}{R}$ to calculate the value for the p.d. Do not forget to rearrange the equation with great care. Many candidates lose marks for poor algebraic skills.

---

(6) (a) $\text{resistance} = \dfrac{\text{voltage}}{\text{current}}$

(b) $R = \dfrac{\rho \ell}{A}$

$$\rho = \frac{RA}{\ell}$$

$$\therefore \rho \rightarrow [\Omega \times \text{m}^2/\text{m}] \rightarrow [\Omega\,\text{m}]$$

(c) (i) $P = \dfrac{V^2}{R}$

$$R = \frac{V^2}{P} = \frac{12^2}{60}$$

$$R = 2.4\,\Omega$$

(ii) $A = \pi r^2 = \pi (3.2 \times 10^{-4})^2 = 3.217 \times 10^{-7}\,\text{m}^2$

$$\ell = \frac{RA}{\rho}$$

$$\ell = \frac{2.4 \times 3.217 \times 10^{-7}}{4.2 \times 10^{-7}}$$

$$\ell = 1.84 \approx 1.8\,\text{m}$$

The length of the wire is too long for the size of the bulb. It must therefore be coiled.

(7) (a) The resistance **decreases** as the incident intensity of light increases.

(b) $R_T = 1200 + 820 = 2020\,\Omega$

$$I = \frac{V}{R} = \frac{5.0}{2020} = 2.475 \times 10^{-3}\,A$$

$V = IR$

$V = 2.475 \times 10^{-3} \times 1200$

$V = 2.97 \approx 3.0\,V$

---

examiner's tip

The procedure above is simple and shows a good understanding of a series circuit.

The potential difference, V across a resistor is also given by

$$V = \frac{R_2}{(R_1 + R_2)} \times V_0$$

where V_0 is the total p.d. across the potential divider circuit and R_1 and R_2 are the resistances of the resistors. The output is taken across the resistor with resistance R_2. This equation gives

$$V = \frac{1200}{1200 + 820} \times 5.0$$

$$V = 2.97 \approx 3.0\,V$$

(c) For no light, the resistance of LDR is very large, therefore the p.d. across 1.2 kΩ resistor is very small (almost zero).

Curve 'tending to' $V \approx 5.0\,V$ in brighter light conditions.

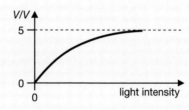

(8) (a) total resistance $= 0.8 + 4.7 + (\frac{4.7}{2}) = 7.85\,\Omega$

$$I = \frac{V}{R} = \frac{1.2}{7.85} = 0.153\,\text{A} \approx 0.15\,\text{A}$$

> **examiner's tip**
>
> The total resistance of the two resistors in parallel may be determined using
> $$\frac{1}{R_T} = \frac{1}{R_1} + \frac{1}{R_2}.$$
> A quicker route would be to use the idea that for two resistors connected in parallel, with each having a resistance R, the total resistance is equal to R/2. Do not forget to include the resistance of the internal resistor in the calculation for the total resistance of the circuit.

(b) external load $= 4.7 + \left(\frac{4.7}{2}\right) = 7.05\,\Omega$

terminal p.d. = p.d. across the external resistance or load
terminal p.d. $= IR = 0.153 \times 7.05 = 1.08\,\text{V}$

(c) fraction of power lost within battery $= \dfrac{I^2 r}{IE}$

where E is the e.m.f. of the cell and r is the internal resistance of the battery.

fraction $= \dfrac{Ir}{E}$

fraction $= \dfrac{0.8 \times 0.153}{1.2} \approx 0.10$

> **examiner's tip**
>
> About 10% of the power delivered by the battery is 'lost' as heat within the battery due to its internal resistance.

(9) (a) $R = 15 \times 60 = 900\,\Omega$

(b) Resistance of wires between **A** and the rod $= 2 \times (30 \times 15) = 900\,\Omega$
Total resistance $= 900 + 300 = 1200\,\Omega$

$$I = \frac{V}{R} = \frac{6.0}{1200} = 0.005\,\text{A}$$

$$V = IR = 0.005 \times 300 = 1.5\,\text{V}$$

> **examiner's tip**
>
> You can also calculate the p.d. across the 300 Ω resistor by using the potential divider equation as shown below:
> $$V = \frac{R_2}{R_1 + R_2} \times V_0 = \frac{300}{900 + 300} \times 6.0 = 1.5\,\text{V}$$

(c) The rod rolling on the resistance wires is equivalent to a variable resistor. As the rod moves from **A** to **B**, the total resistance of the circuit increases and therefore the p.d. across the 300 Ω decreases as the resistance wires take a greater share of the potential difference.

Electricity

(10) (a) **Kirchhoff's first law**:
The sum of the currents entering a junction is equal to the sum of the currents leaving the junction.

Kirchhoff's second law:
The sum of the electromotive forces in a loop is equal to the sum of the potential differences in that loop.

(b) **(i)** Applying Kirchhoff's first law to the junction **X**, we have
$I = 0.30 + 0.15 = 0.45\,A$

(ii) Applying Kirchhoff's second law to the 'loop' with the 5.25 V cell, 10 Ω resistor and the resistor of resistance R, we have
sum of e.m.f.s = sum of p.d.s
$5.25 = (0.15 \times R) + (0.45 \times 10)$
$R = \dfrac{5.25 - 4.50}{0.15} = 5.0\,\Omega$

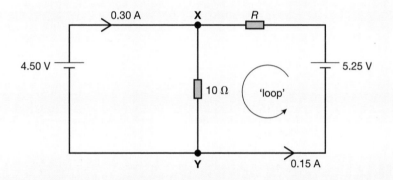

examiner's tip   You can also get the answer by applying Kirchhoff's second law to the 'outermost loop' with the two cells and the resistor of resistance *R*.

Electricity

33

Chapter 3

Thermal physics and radioactivity

Questions with model answers

C grade candidate – mark scored 6/10

In a data book, the following information is given about lithium.

Symbol: Li

Atomic (proton) number: 3

Isotopes: Lithium-6 (stable)

Lithium-7 (stable)

Lithium-8 (emits β-particles)

? For help see Revise AS Study Guide pages 99, 100 and 101

Examiner's Commentary

The candidate is correct to refer to nuclei and not to atoms. An atom consists of a positive nucleus, but it also consists of the negatively charged electrons.

1 What are **isotopes**? [2]

Isotopes are nuclei with the same number of protons ✔ *but different number of neutrons.* ✔

There are two marks and the candidate has given just one response. It would have been advantageous to state that 'beta-particles are electrons that carry negative charge away from the nucleus.'

2 For the lithium-7 isotope, state the number of

(a) protons

The nucleus has three protons. ✔

(b) neutrons [2]

N = A − Z
N = 7 − 3 = 4   There are four neutrons. ✔

3 The lithium-8 isotope emits beta-particles.

(a) State the nature of beta-particles. [2]

β particles carry negative charge. ✔

(b) With reference to a nuclear decay equation, state and explain the change taking place within the nucleus of lithium-8 when it emits a beta-particle. [4]

The β particle carries energy away from the nucleus making it more stable. ✔

This is a very brief answer. The candidate was awarded one mark for appreciating that the decay is an attempt by the nucleus to become more stable. However, there is no nuclear decay equation. If this were to be done, then it would have been a prompt for the candidate to think in terms of the nucleon and the proton numbers. The decay equation for the nucleus is
$$^{8}_{3}\text{Li} \rightarrow {}^{0}_{-1}\text{e} + {}^{8}_{4}\text{Be}.$$
The nucleon number remains the same, but there is one **extra** proton within the nucleus. This could only have happened if a neutron inside the nucleus transforms into a proton and an electron. The electron is the beta-particle that is emitted by the unstable nucleus of lithium-8. Within the nucleus, the following change takes place
neutron → proton + electron + anti-neutrino
$$^{1}_{0}\text{n} \rightarrow {}^{1}_{1}\text{p} + {}^{0}_{-1}\text{e} + \bar{v}_e$$

GRADE BOOSTER

Keep an eye on the marks available. If there are four marks available then an examiner will look for an equal number of valid responses to the question. In **3(b)**, the candidate has only listed one item for a question with four marks.

1 Write the equation of state for an ideal gas.   [1]

$$PV = nRT \quad ✔$$

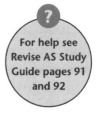

For help see
Revise AS Study
Guide pages 91
and 92

2 The diagram shows a metal cylinder in which some gas is trapped by a piston.

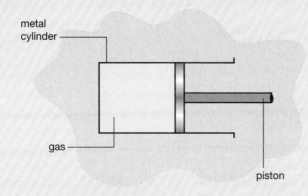

metal
cylinder

gas

piston

(a) The volume of the trapped gas is 1.7×10^{-3} m³.
At a temperature of 210°C, the pressure exerted by the gas is 3.5×10^5 Pa. Calculate the number of moles of gas within the cylinder. You may assume that the trapped gas behaves like an ideal gas.
Data: $R = 8.31$ J mol^{-1} K^{-1} [3]

$$n = \frac{PV}{RT}$$

$$T = 273 + 210 = 483 \text{ K} \qquad ✔$$

$$n = \frac{3.5 \times 10^5 \times 1.7 \times 10^{-3}}{8.31 \times 483} \qquad ✔$$

$$n = 0.15 \text{ mol} \qquad ✔$$

> The most common mistake made by candidates is to use the temperature in celsius. In the equation above, T is the thermodynamic temperature. There was one mark reserved for the correct conversion of the gas temperature.

(b) The gas within the cylinder expands. The volume occupied by the gas **increases** by 30%, but the pressure is maintained at 3.5×10^5 Pa. Calculate the final temperature of the gas.   [2]

$$\frac{PV}{T} = \text{constant}$$

$$\frac{V}{T} = \text{constant or } V \propto T \qquad ✔$$

If the volume increases, so does the temperature by the same factor.
Therefore $T = 483 \times 1.3 = 628$ K   ✔

1 **(a)** State Boyle's law. [2]

(b) A diver working at a depth of 15 m in the sea releases a rubber balloon to mark his position. At this depth, the volume of the balloon is 4.0×10^{-3} m³ and the pressure exerted by the gas within the balloon is 2.5×10^5 Pa. The temperature of the sea water is 18°C.

    (i) Show that the amount of gas within the balloon is 0.41 mol.
Data: $R = 8.31$ J mol^{-1} K^{-1} [3]

    (ii) Calculate the number of gas molecules within the balloon.
Data: $N_A = 6.02 \times 10^{23}$ mol^{-1} [2]

    (iii) The volume of the balloon increases as it rises towards the surface.
At the water surface, the pressure exerted by the gas within the balloon decreases to 1.0×10^5 Pa. Calculate the volume of the balloon at the surface. [2]

[Total: 9]

---

2 **(a)** Explain in molecular terms the origin of pressure within a container. [4]

(b) For an ideal gas, the mean translational kinetic energy E_k of a molecule is given by

$$E_k = \tfrac{3}{2}kT$$

Define the symbols k and T. [2]

(c) The surface temperature of a star is 5800 K. On its surface, protons move randomly and behave like the molecules of an ideal gas. For these surface protons, calculate

    (i) the mean translational kinetic energy of each proton,
Data: $k = 1.38 \times 10^{-23}$ J K^{-1} [2]

    (ii) the root-mean-square (r.m.s.) speed of the protons.
Data: mass of proton, $m_p = 1.7 \times 10^{-27}$ kg [3]

(d) Explain how the answer to **(c)(ii)** would change if the particles on the surface of the star were helium nuclei. [2]

[Total: 13]

Answers on pages 41–45 Answers on pages 41–45 Answers on pages 41–45

3 **(a)** Define specific heat capacity of a substance. [1]

(b) A 1.0 kW electric kettle contains 450 g of water at 15°C. The specific heat capacity of the water is $4.2 \times 10^3 \, \text{J kg}^{-1} \, \text{K}^{-1}$.

  (i) Calculate the energy supplied by the heating element of the kettle to raise the temperature of the water to 100°C. State any assumption made. [3]

  (ii) How long would it take to raise the temperature of the water to 100°C? [2]

  (iii) The graph below shows the actual variation of the temperature θ of the water inside the electric kettle with time t.
  (The kettle is switched on at time $t = 0$ s)

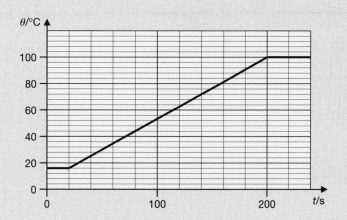

  1. Suggest why the temperature of the water remains constant for the first 20 s. [1]

  2. What is happening to the energy supplied to the water after 200 s? [1]

[Total: 8]

4 **(a)** A radioactive material emits alpha (α) particles. What is the nature of alpha-particles? [2]

(b) The radioactive source used in many domestic smoke alarms is americium-241. The source is housed within a plastic case. The isotopes of americium-241 have a half-life of 460 years. In each decay of ^{241}Am nucleus, an alpha-particle of kinetic energy 5.4 MeV is released.

For one particular source in a smoke alarm, the activity is 3.5×10^3 Bq. Calculate

   (i) the decay constant, λ, for the isotopes of ^{241}Am, [2]

   (ii) the number of ^{241}Am nuclei in the source, [3]

   (iii) the rate of energy released by the source.
   Data: 1 MeV $= 1.6 \times 10^{-13}$ J. [3]

(c) Why is it sensible to use an americium-241 source within the domestic smoke detector rather than a source that emits either beta-particles or γ-rays? [2]

[Total: 12]

---

5 **(a)** Explain what is meant by the decay constant of a nuclide. [1]

(b) Define the half-life of a nuclide. [2]

(c) Iodine-131 has a half-life of about 8 days. Calculate

   (i) the fraction of iodine nuclei left in a sample after 32 days, [2]

   (ii) the fraction of iodine nuclei that would have decayed in a sample after 16 days. [2]

[Total: 7]

6 (a) Explain what is meant by the absolute zero of temperature. [2]

(b) The diagram shows two insulated gas cylinders connected together by a small tube of negligible volume.

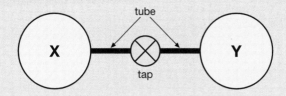

Each of the cylinders has an internal volume of $1.6 \times 10^{-2}\,\text{m}^3$ and contains an ideal gas. The gas in cylinder **X** exerts a pressure of $1.0 \times 10^5\,\text{Pa}$ and the gas in cylinder **Y** exerts a pressure of $2.1 \times 10^5\,\text{Pa}$. The temperature within each cylinder is $27°C$.
Data: $R = 8.31\,\text{J}\,\text{mol}^{-1}\,\text{K}^{-1}$

   (i) Calculate the amount, in mol, of the gas in cylinder **Y**. [3]

   (ii) The tap is slowly opened and the gases in both cylinders are allowed to mix. The gases do not react chemically and their temperature remains constant. Determine the new pressure of the gas within the cylinders. [3]

[Total: 8]

---

7 (a) Radioactive decay of nuclei is a *random* and a *spontaneous* event. Explain what is meant by the terms in italics. [2]

(b) A radioactive material contains an isotope of radon ($^{220}_{86}\text{Rn}$). When a nucleus of radon-220 decays, it emits an alpha-particle and changes into a nucleus of polonium (Po).

   (i) State the number of neutrons within the nucleus of radon-220. [1]

   (ii) Name two quantities that are conserved in the decay of radon-220. [2]

   (iii) Write a nuclear equation to represent the decay of radon-220 by an emission of an alpha-particle. [2]

[Total: 7]

8 **(a)** Write a defining equation for the specific heat capacity of a substance. State the meaning of any symbols you have used. [2]

(b) A metal of mass 200 g is heated to a temperature of 570°C. The metal is quickly immersed into a beaker of water. The initial temperature of the water is 20°C and it has a mass of 300 g. Calculate the final temperature of the water. (You may assume that there is no loss of heat to the surroundings.)

Data: specific heat capacity of water = 4200 J kg^{-1} K^{-1}

specific heat capacity of the metal = 800 J kg^{-1} K^{-1} [3]

[Total: 5]

Answers on pages 41–45 **Answers** on pages 41–45 **Answers** on pages 41–45

(1) (a) For a fixed amount of ideal gas at constant temperature, the pressure exerted by the gas is inversely proportional to its volume.

(b) **(i)** $T = 273 + 18 = 291\,K$

$PV = nRT$

$2.5 \times 10^5 \times 4.0 \times 10^{-3} = n \times 8.31 \times 291$

$n = 0.414\,mol$

examiner's tip

It is very important that the temperature is converted into kelvin. Using 18°C would lead to the wrong answer.

(ii) number $= n \times N_A$

number $= 0.414 \times 6.02 \times 10^{23}$

number $= 2.49 \times 10^{23}$

$\approx 2.5 \times 10^{23}$ molecules

(iii) $PV = $ constant

(since the temperature is constant)

$2.5 \times 10^5 \times 4.0 \times 10^{-3} = 1.0 \times 10^5\,V$

$V = 1.0 \times 10^{-2}\,m^3$

examiner's tip

Another route to the correct answer would be to use

$$PV = nRT$$

together with the value of n from **(b)(i)**. There is not much to choose from between this and the technique shown.

(2) (a) Any *four* from:

The molecules are moving about in a random manner.

Each molecule collides with the container wall, resulting in a **change** in velocity (or momentum).

According to Newton's second law, there is a tiny force exerted on the molecule by the wall.

There is an equal but opposite force exerted on the wall by the molecule (Newton's third law).

There are numerous molecular collisions with the walls giving rise to a larger force on the wall.

The pressure P on the wall is given by

$$P = \frac{\text{force on wall}}{\text{area of wall}}.$$

examiner's tip

There are four marks for this question. It is important that you do not repeat the same points over and over again. Before answering the question, it is advisable to make a quick list of four distinct points relevant to the question.

(b) k is the Boltzmann constant.

T is the thermodynamic temperature in kelvin.

Thermal physics and radioactivity

(c) (i) $E_k = \frac{3}{2}kT$

$E_k = \frac{3}{2} \times 1.38 \times 10^{-23} \times 5800$

$E_k = 1.2 \times 10^{-19}$ J

(ii) $E_k = \frac{1}{2}m\langle c^2 \rangle$

$\langle c^2 \rangle = \dfrac{2 \times 1.2 \times 10^{-19}}{1.7 \times 10^{-27}}$

r.m.s. speed $= \sqrt{1.412 \times 10^8}$

r.m.s. speed $= 1.19 \times 10^4 \approx 1.2 \times 10^4 \, \text{m s}^{-1}$

(d) The mean translational kinetic energy is the same.

The more **massive** helium nuclei would move much slower.

(3) (a) The specific heat capacity of a substance is the energy required to change the temperature of a unit mass of the substance by 1 K (or 1°C).

(b) (i) The assumption is that there is no loss of heat to the surroundings.

$\Delta E = mc\Delta\theta$

$\Delta E = 0.45 \times 4.2 \times 10^3 \times (100 - 15)$

$\Delta E = 1.61 \times 10^5 \approx 1.6 \times 10^5$ J

(ii) $\Delta t = \dfrac{\Delta E}{P}$

$\Delta t = \dfrac{1.61 \times 10^5}{1.0 \times 10^3}$

$\Delta t = 161 \approx 160$ s

(iii) 1. The heating element of the kettle is being heated instead of the water.

2. The water is boiling. The energy supplied to the water is used to break molecular bonds as the water changes state into steam.

Thermal physics and radioactivity

(4) (a) Any *two* from:
Alpha-particles are:
helium **nuclei**,
positively charged
and consist of two protons and two neutrons.

(b) (i) 1 year $= 365 \times 24 \times 3600 = 3.154 \times 10^7 \text{ s}$

$$\lambda = \frac{\ln(2)}{T_{\frac{1}{2}}} \approx \frac{0.693}{T_{\frac{1}{2}}}$$

$$\lambda = \frac{0.693}{460 \times 3.154 \times 10^7}$$

$$\lambda = 4.78 \times 10^{-11} \approx 4.8 \times 10^{-11} \text{ s}^{-1}$$

examiner's tip It is important that the answer is given in S.I. units. It is therefore sensible to convert the half-life into seconds.

(ii) The decay equation is

$$\frac{\Delta N}{\Delta t} = -\lambda N \text{ or } A = -\lambda N$$

where A is the rate of decay of nuclei (or the activity of the source).

$$N = \frac{A}{\lambda} \quad \text{(Ignore the minus sign)}$$

$$N = \frac{3.5 \times 10^3}{4.78 \times 10^{-11}}$$

$$N = 7.32 \times 10^{13} \approx 7.3 \times 10^{13} \text{ nuclei}$$

(iii) K.E. of each α-particle $= 5.4 \text{ MeV}$
K.E. of each α-particle $= 5.4 \times 1.6 \times 10^{-13} = 8.64 \times 10^{-13} \text{ J}$

Power $=$ rate of energy release
Power $=$ activity $\times$ energy of each α-particle
$P = 3.5 \times 10^3 \times 8.64 \times 10^{-13}$
$P = 3.02 \times 10^{-9} \approx 3.0 \times 10^{-9} \text{ J s}^{-1} \text{ (W)}$

examiner's tip This question requires a good understanding of the term **activity** which represents the number of emissions from the source per unit time. Since power is the amount of energy released per unit time, it follows that:
power = activity × energy of each α-particle.

(c) It is safer.
This is because the α-particles have a short range in air and will not travel outside the plastic case of the alarm but β-particles and γ-rays can.

Thermal physics and radioactivity

(5) (a) The decay constant of a nuclide is the probability of decay of the nuclide per unit time.

(b) The half-life of a nuclide is the **mean** time taken for half the nuclei in a sample to decay.

(c) (i) There are four half-lives in the time interval of 32 days.

$$\text{fraction of active nuclei left} = (\frac{1}{2})^4 = \frac{1}{16}$$

(ii) There are two half-lives in a time interval of 16 days.

$$\text{fraction of active nuclei \textbf{left}} = (\frac{1}{2})^2 = \frac{1}{4}$$

$$\text{fraction of nuclei \textbf{decayed}} = 1 - \frac{1}{4} = \frac{3}{4}$$

examiner's tip This question requires careful reading. The number of nuclei that have decayed is the difference between the original number and the number of nuclei left in the sample.

(6) (a) At absolute zero of temperature, 0 K, the mean kinetic energy of the atoms or molecules within a sample is equal to zero.

examiner's tip It is important to make reference to either atoms or molecules. Do not use 'particle' because is a bit vague.

(b) (i) $PV = nRT$

$$n = \frac{PV}{RT} = \frac{2.1 \times 10^5 \times 1.6 \times 10^{-2}}{8.31 \times (273 + 27)}$$

$$n = 1.348 \approx 1.3$$

examiner's tip You must change the temperature into kelvin.

(ii) Before the tap is opened, the number of mols of gas within cylinder **X** is given by

$$n = \frac{PV}{RT} = \frac{1.0 \times 10^5 \times 1.6 \times 10^{-2}}{8.31 \times (273 + 27)} = 0.642$$

After the mixing,
$n = 0.642 + 1.348 = 1.99$
The final pressure is P.
$PV = nRT$
$P \times (2 \times 1.6 \times 10^{-2}) = 1.99 \times 8.31 \times 300$

$$P = \frac{1.348 \times 8.31 \times 300}{3.2 \times 10^{-2}} = 1.55 \times 10^5 \, \text{Pa} \approx 1.6 \times 10^5 \, \text{Pa}$$

(7) (a) **Random** means that it is impossible to predict which particular nucleus will decay at a particular time. However, for a large number of nuclei, we can predict the number of nuclei in the sample that will decay in a small interval of time. Each nucleus has the same probability of decay per unit time.

Spontaneous means that the decay of the nuclei is unaffected by chemical reactions or external factors such as pressure and temperature. The decay of the nuclei is not affected by the presence of the other nuclei in the sample.

(b) **(i)** number of neutrons $= 220 - 86 = 134$

(ii) Any two from:
charge (or proton number), nucleon number, momentum and mass & energy

(iii) $^{220}_{86}\text{Rn} \rightarrow {}^{4}_{2}\text{He} + {}^{216}_{84}\text{Po}$

---

(8) (a) $c = \dfrac{Q}{m\Delta T}$

where c = specific heat capacity of the substance
Q = heat supplied to the substance
m = mass of the substance
ΔT = change in the temperature of the substance

(b) heat transferred **from** the metal = heat transferred **to** the water
θ = final temperature of the water
$0.200 \times 800 \times (570 - \theta) = 0.300 \times 4200 \times (\theta - 20)$
$91\,200 - 160\theta = 1260\theta - 25\,200$
$(1260 + 160)\theta = 91\,200 + 25\,200$

$\theta = \dfrac{116\,400}{1420} = 81.97°\text{C} \approx 82°\text{C}$

examiner's tip | You have to set up an equation for the unknown temperature and then solve it. In this question you do need good algebraic skills.

Questions with model answers

C grade candidate – mark scored 6/9

1 Explain what is meant by a photon. [1]

A photon is how energy is carried by an electromagnetic wave.

It is a quantum (or packet) of energy. ✔

A photon travels at the speed of light.

? For help see Revise AS Study Guide pages 130 and 131

The candidate secured the second mark because the correct prefix of the 'milli' was inserted in front of the unit for power. It is always safer to substitute and write answers in standard form.

2 A light-emitting diode (LED) emits visible light when it conducts. For one particular LED, it **just** starts to emit red light when the p.d. across it is 1.8 V and the current through it is 1.2 mA. The light emitted by the LED has a wavelength of 6.8×10^{-7} m.

(a) What is the input electrical power to the LED? [2]

$P = VI$

$P = 1.8 \times 1.2$ ✔

$P = 2.16 \text{ mW}$ ✔

(b) Calculate the energy of each photon of red light from the LED.

Data: $h = 6.63 \times 10^{-34}$ J s

$c = 3.0 \times 10^{8}$ m s^{-1} [3]

$E = hf$ ✔

$E = \dfrac{h\lambda}{c}$ ✘

$E = \dfrac{6.63 \times 10^{-34} \times 6.8 \times 10^{-7}}{3.0 \times 10^{8}} = 1.5 \times 10^{-48} \text{ J}$ ✘

The first mark was awarded for correctly recalling the equation for the energy of a photon. The candidate's second step is wrong. Since
$$c = f\lambda \text{ and } E = hf$$
$$\text{then } E = \frac{hc}{\lambda}.$$
Since the incorrect equation was used by the candidate, no further marks could be awarded. The correct answer is 2.9×10^{-19} J.

(c) Use your answers to **(a)** and **(b)** to estimate the rate at which photons are emitted from the LED. State any assumption made. [3]

I am going to assume that all the electrical power goes into producing light. ✔

$P = \dfrac{\Delta E}{\Delta t}$

When $t = 1$ s $\qquad \Delta E = 2.16 \text{ mJ}$

Number of photons $= \dfrac{2.16}{1.5 \times 10^{-48}}$ ✘

Number of photons $= 1.4 \times 10^{48}$ per second. ✔ (error carried forward)

The candidate has used the wrong answer from **(b)**. This, by itself would have been alright here. Examiners try not to penalise twice for the same error. However, another mistake in the form of the missing 10^{-3} factor for the power has slipped in. The examiner has deducted one mark for this error, but subsequent marks have been awarded. The correct answer for the rate of photon emission from the LED would have been 7.4×10^{15} s^{-1}.

GRADE BOOSTER Algebra is the language used in physics. Make sure you can recall and rearrange equations. This candidate lost two valuable marks in **2(b)** because of poor algebraic skills. An extra two marks would have given a grade A to this candidate.

A grade candidate – mark scored 5/5

1 Explain what is meant by a **transverse** wave. [1]

A transverse wave has oscillations that are at right angles to the wave velocity. ✔

For help see Revise AS Study Guide pages 117 and 120

2 According to a student

'Light reflected from the surface of water is **plane polarised**.'

(a) State what is meant by **plane polarised**. [1]

This is when a wave (transverse) has oscillations in only one plane. ✔

(b) Name a wave, other than visible light, that can be plane polarised. [1]

All electromagnetic waves can be polarised. ✔

All the following waves can be polarised:

γ-rays, X-rays, ultra-violet, infra-red, microwaves and radio waves.

This is a superb answer. The response is typical from a grade A candidate. The candidate has secured the mark by stating that all electromagnetic waves can be polarised, but has elaborated a bit more, just to be on the safe side.

(c) Outline an experiment to assess the validity of the student's statement. [2]

I would look at the reflected light from the water through a polarising filter. ✔

I would rotate the filter. If the reflected light was polarised, then the intensity of the light seen through the filter would vary. ✔

Since the reflected light from the water is thought to be plane polarised, it is possible to use just one polarising filter rotated in front of the eye.

Wave properties and electromagnetic waves

1 Calcium has a line spectrum, which includes the spectral line at a wavelength of 393 nm.
Data: $c = 3.0 \times 10^8 \, \text{m s}^{-1}$

(a) Calculate the frequency of this line. [2]

(b) To which region of the electromagnetic spectrum does this line belong? [1]

(c) What is a line spectrum? [1]

Edexcel June 2003

[Total: 4]

2 **(a)** The diagram shows a ray of light within a semi-circular block of glass making an angle of incidence equal to the **critical angle**, c.

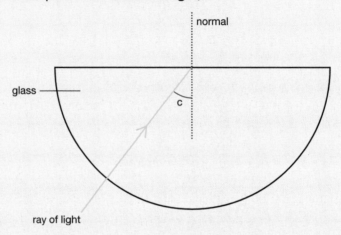

(i) Complete the diagram above to show what happens to the ray of light. [2]

(ii) The critical angle, c for the glass is 42°. Calculate the refractive index for the glass. (You may assume that air has a refractive index of 1.00.) [2]

(b) The diagram below shows the cross-section through an optical cable made of glass of refractive index 1.52. A ray of light incident at an angle of 20° to the glass–air interface is internally reflected. The total length of the optical cable is 1.20 km.

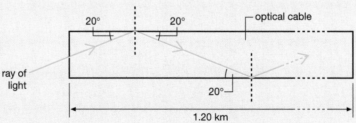

(i) Calculate the speed of light in the optical cable.
Data: $c = 3.0 \times 10^8 \, \text{m s}^{-1}$ [2]

(ii) For a ray of light incident at an angle of 20° to the glass–air interface, calculate the time taken for it to travel the total length of the cable. [3]

(iii) State how the answer to **(b)(ii)** would change for a ray of light travelling along the axis of the cable. [1]

(iv) Suggest why it may be sensible for an optical cable to have a very small diameter when transmitting digital information. [1]

[Total: 11]

Answers on pages 53–58 Answers on pages 53–58 Answers on pages 53–58

3 **(a)** Outline two properties of a progressive wave. [2]

(b) The diagram shows two identical loudspeakers emitting sound when connected to a signal generator set at 1.2 kHz.

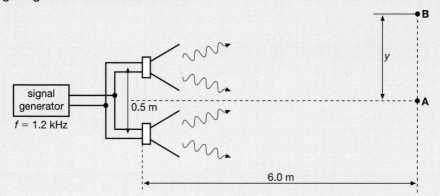

(i) Calculate the wavelength of sound emitted from each loudspeaker.
Data: speed of sound in air = 340 m s^{-1} [2]

(ii) A student is listening to the sound from **both** loudspeakers.
When at **A**, a very loud sound is heard. As the student slowly moves towards **B**, the intensity of sound gradually decreases. When at **B**, virtually no sound is heard. Explain why a loud sound is heard at point **A** and virtually none at the adjacent point **B**. [2]

(iii) Calculate the distance y between **A** and **B**. [3]

(iv) State and explain how the answer to **(b)(iii)** would change if the frequency of sound were to be doubled. (You are not expected to do any further calculations.) [2]

[Total: 11]

---

4 **(a)** Outline some of the main properties of electromagnetic waves. Name one principal region of the electromagnetic spectrum and suggest a practical application for the waves. [5]

(b) A gold-leaf electroscope consists of a metal cap and a thin strip of gold foil attached at the end of the metal stem. The diagram shows a negatively charged electroscope, with the gold-leaf diverged.

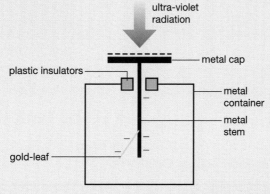

The metal cap of the electroscope is made of zinc. When the cap is exposed to a weak ultra-violet source, the gold-leaf starts to collapse.
After some time, it shows no divergence.

Outline the phenomenon of the photoelectric effect and use the ideas developed to explain why the divergence of the gold-leaf decreases with time. [5]

[Total: 10]

Wave properties and electromagnetic waves

5 **(a)** State **one** common feature of all electromagnetic waves travelling through a vacuum. [1]

(b) Two radio transmitters broadcast signals at frequencies of 198 kHz and 102 MHz.

    (i) State which of the two signals has a longer wavelength. [1]

    (ii) Calculate the ratio: [2]

$$\frac{\text{wavelength of 198 kHz signal}}{\text{wavelength of 102 MHz signal}}$$

    (iii) The diagram shows a small community in a hilly region receiving signals from the two transmitters.

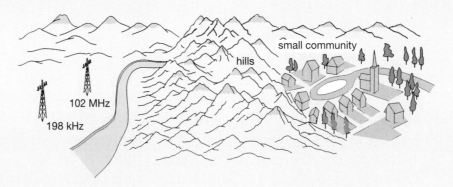

        Explain why there are no problems receiving the signals at the longer wavelength. [2]

[Total: 6]

---

6 **(a)** Suggest **two** differences between a standing (or stationary) wave and a progressive wave. [2]

(b) A loudspeaker, connected to a signal generator set at 2.0 kHz, is placed in front of a smooth flat vertical wall. This is illustrated in the diagram below.

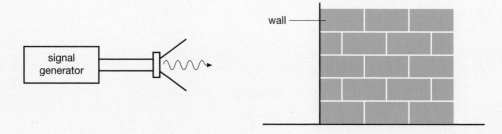

    (i) Explain how a standing wave is created between the loudspeaker and the wall. [2]

    (ii) A small microphone moved from the loudspeaker towards the wall detects a series of maxima and minima. Calculate

        1. the wavelength λ of the sound emitted by the loudspeaker,
        Data: speed of sound = 340 m s^{-1} [2]

        2. the distance between successive maxima. [2]

    (iii) State and explain how your answer to **(b)(ii)2** would change if the frequency of sound is increased. [2]

[Total: 10]

7 The diagram shows a long horizontal plastic tube containing some fine powder.
One end of the tube is closed and a loudspeaker is positioned at the other end.
The loudspeaker is connected to a signal generator.

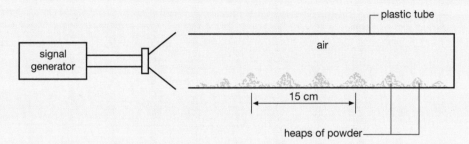

At a particular frequency, a standing wave is created in the air column within the tube.
The powder within the tube forms heaps at the **nodes**.

(a) State why the powder forms heaps at the **nodes**. [1]

(b) Determine the wavelength λ of the sound wave. [3]

(c) Calculate the frequency of the sound produced by the loudspeaker.
Data: speed of sound = 340 m s^{-1} [2]

(d) Suggest why the powder disperses when the frequency of sound is altered slightly. [1]

[Total: 7]

---

8 **(a)** Explain what is meant by the work function energy of a metal. [1]

(b) A photon having energy less than the work function energy of the metal
interacts with a single electron in the metal. Explain what happens to the energy
of the electron. [2]

(c) The work function energy of a particular metal is 2.5 eV. Calculate

    (i) the threshold frequency,
    Data: 1 eV = 1.6 × 10^{-19} J [2]

    (ii) the maximum kinetic energy of an emitted electron when the metal is exposed
    to electromagnetic radiation of wavelength 3.6 × 10^{-7} m.
    Data: h = 6.63 × 10^{-34} J s [3]

[Total: 8]

Wave properties and electromagnetic waves

9 Electromagnetic radiation is incident on a metal surface. The results from an experiment are shown on the graph below. The maximum kinetic energy of the photoelectrons is E_k and f is the frequency of the incident electromagnetic radiation.

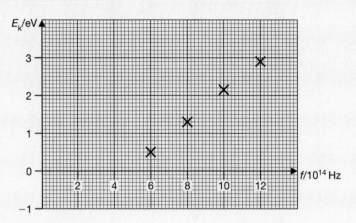

(a) Draw a line of best-fit. [1]

(b) State why there are no data points for negative values for E_k. [1]

(c) Use the graph to determine the threshold frequency f_0 of the metal. [1]

(d) Calculate the work function energy ϕ of the metal in joules.
Data: $h = 6.63 \times 10^{-34}\,\text{J s}$ [3]

(e) Explain why the gradient of the line is equal to the Planck constant h. [2]

[Total: 8]

(1) (a) $c = f\lambda$

$$f = \frac{3.0 \times 10^8}{393 \times 10^{-9}} \approx 7.63 \times 10^{14}\,\text{Hz}$$

(b) Ultra-violet region of the spectrum.

(c) A few distinct wavelengths of electromagnetic radiation that are either emitted or absorbed by the atoms. These wavelengths are characteristic to the type of atom.

---

(2) (a) **(i)** The ray is refracted at 90°.
There is a weak reflection within the glass block.

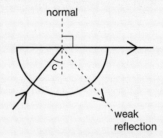

(ii) $n = \dfrac{1}{\sin c}$

where n is the refractive index and c is the critical angle within the glass.

$$n = \frac{1}{\sin 42°}$$

$$n = 1.494 \approx 1.49$$

> **examiner's tip**
>
> According to Snell's law $\qquad \dfrac{\sin i}{\sin r} = {}_1 n_2$
>
> As the ray of light travels towards the glass–air interface, an incidence angle of 42° within **glass** gives a refracted angle of 90° in **air**. Therefore
>
> $$\frac{\sin 42°}{\sin 90°} = {}_{\text{glass}} n_{\text{air}}$$
>
> But $\qquad\qquad\qquad {}_{\text{air}} n_{\text{glass}} = \dfrac{1}{{}_{\text{glass}} n_{\text{air}}}$
>
> $$\therefore\ {}_{\text{air}} n_{\text{glass}} = \frac{1}{\sin 42°} \approx 1.49$$

(b) **(i)** $n = \dfrac{c_v}{c_m}$

$$c_m = \frac{3.00 \times 10^8}{1.52}$$

$$c_m = 1.974 \times 10^8 \approx 1.97 \times 10^8\,\text{m s}^{-1}$$

(ii) $\text{distance} = \dfrac{1.2 \times 10^3}{\cos 20°} = 1.277 \times 10^3\,\text{m}$

$$\text{time} = \frac{\text{distance}}{\text{velocity}}$$

$$t = \frac{1.277 \times 10^3}{1.974 \times 10^8}$$

$$t \approx 6.47 \times 10^{-6}\,\text{s}\ (6.5\,\mu s)$$

The actual 'zigzag' path of the reflected ray can be stretched out as shown below.

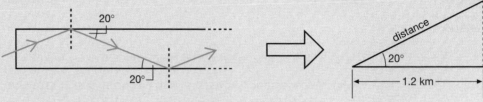

The total distance travelled by the internally reflected ray is given by

$$\cos 20° = \frac{1.2 \times 10^3}{\text{distance}}$$

In order to determine the time taken, you must use the actual velocity of the ray within the optical cable. To use a velocity of $3.00 \times 10^8 \, \text{m s}^{-1}$ would be incorrect.

(iii) The time taken would be shorter.

The reason for the shorter time is because the distance travelled by the ray would be less than the longer 'zigzag' path.

(iv) There would be reduced signal 'smearing'.

(3) (a) Any *two* from:

A progressive wave travels through space carrying energy.
It carries energy through space via vibrations (or oscillations).
It has wavelength, frequency and wave velocity.
The wave velocity, v, is given by $v = f\lambda$.

(b) (i) $v = f\lambda$

$$\lambda = \frac{340}{1.2 \times 10^3}$$

$$\lambda = 0.283 \approx 0.28 \, \text{m}$$

(ii) At **A**, the waves from the loudspeakers interfere **constructively**.
At **B**, the waves from the loudspeakers interfere **destructively**.

(iii) $x = \dfrac{\lambda D}{a}$

$$x = \frac{0.283 \times 6.0}{0.5} = 3.40 \, \text{m}$$

$$y = \frac{x}{2}$$

$$y = 1.70 \approx 1.7 \, \text{m}$$

The separation x between neighbouring maximum (or minimum) signals is given by the expression $x = \dfrac{\lambda D}{a}$.

In the question, the required distance is between the maximum signal at A and the adjacent minimum signal at B. Therefore, the distance y will be half that of x.

(iv) The wavelength of sound would be halved.
Therefore, y will be reduced by a factor of two.

examiner's tip There are two main ideas here. They are $v = f\lambda$ and $x = \dfrac{\lambda D}{a}$.

$$\therefore x = \frac{Dv}{af}$$

$$\text{or } x \propto \frac{1}{f}.$$

The separation x, and hence y, will be **halved** when the frequency of the signal is **doubled**.

(4) (a) Electromagnetic waves are:
Transverse waves that travel through a vacuum at a velocity of $3.0 \times 10^8 \text{ m s}^{-1}$.
E.M. waves consist of oscillating electric and magnetic fields.

One of the principal regions of the electromagnetic spectrum is the infra-red region.
Infra-red radiation is used by television remote controls and security lighting at night.

examiner's tip The principal regions of the electromagnetic spectrum in order of **increasing** wavelength are: γ-rays, X-rays, ultra-violet, visible light, infra-red, microwaves and radio waves.
The practical applications are too numerous to list. Microwaves, for example, are used in microwave ovens and for mobile phone communication.

(b) Negatively charged electrons are removed from the **surface** of zinc by the **photons** interacting with these electrons.
As a result, the divergence of the gold-leaf decreases.
Each photon interacts with a single surface electron.
Energy is conserved, therefore '$hf = \phi + \text{KE}_{max}$'.
Electrons are removed because the energy of each ultra-violet photon is greater than the work function ϕ of the metal.
There is a threshold frequency above which electrons are emitted from the metal surface.
Below the threshold frequency no electrons are emitted, no matter how intense the incident radiation.

(5) (a) All electromagnetic waves travel at $3.0 \times 10^8 \text{ m s}^{-1}$ in a vacuum.

(b) (i) The 198 kHz signal.

(ii) $\text{ratio} = \dfrac{c}{f_1} \div \dfrac{c}{f_2}$

$\text{ratio} = \dfrac{f_2}{f_1} = \dfrac{102 \times 10^6}{198 \times 10^3} = 515$

There is no need to know the actual speed of the waves. Since the speed is a constant, the wavelength λ is related to the frequency f by:

$$\lambda \propto \frac{1}{f}.$$

The ratio is therefore equal to the inverse ratio of the frequencies. You do not require the speed of electromagnetic waves because it cancels out when you substitute to determine the ratio.

(iii) The longer wavelength signal is diffracted by the hills.
This is because the 'gap' between the hills is comparable to the wavelength of these signals.

(6) (a) Energy is 'localised' for a standing wave.
All points between two adjacent nodes oscillate in phase.

(b) (i) The sound is reflected at the wall.
The incident and reflected waves superimpose to produce a standing wave pattern.

(ii) 1. $\lambda = \dfrac{v}{f}$

$\lambda = \dfrac{340}{2000} = 0.17 \text{ m}$

2. The separation between two neighbouring antinodes (or nodes) $= \dfrac{\lambda}{2}$

distance between adjacent maxima $= \dfrac{0.17}{2} = 0.085 \text{ m}$ (8.5 cm)

(iii) The separation between the maxima decreases because the wavelength is shorter. (Note: the separation between neighbouring nodes is equal to $\dfrac{\lambda}{2}$.)

(7) (a) There is no oscillation of air particles at the nodes.

The fine powder will move around and gather at the points where there is locally no oscillations of the air. It is at these points that the powder gathers and forms tiny heaps.

(b) Separation between successive nodes (or antinodes) $= \dfrac{\lambda}{2}$

$\dfrac{3\lambda}{2} = 15 \text{ cm}$ (using the diagram)

$\therefore \lambda = 10 \text{ cm}$

(c) $v = f\lambda$

$f = \dfrac{340}{0.1}$

$f = 3.4 \times 10^3 \text{ Hz}$ (3.4 kHz)

For a standing wave, its frequency is equal to the frequency of the sound from the loudspeaker.

(d) There is no standing wave produced within the plastic tube.

(8) (a) The work function energy of the metal is the **minimum** energy required to liberate an electron from the surface of the metal.

(b) Energy is conserved in the interaction. The photon energy is transferred to the electron as its kinetic energy. This kinetic energy is less than the work function energy of the metal, hence the electron cannot escape. As the electron collides with the atoms of the metal, its kinetic energy is transferred to the atoms as heat.

(c) (i) $hf = \phi + KE_{max}$

At the threshold frequency, f_o, the kinetic energy of the electron is zero.

$$f_o = \frac{\phi}{h} = \frac{2.5 \times 1.6 \times 10^{-19}}{6.63 \times 10^{-34}} = 6.03 \times 10^{14}\,\text{Hz}$$

(ii) $hf = \phi + KE_{max}$

$$KE_{max} = \frac{hc}{\lambda} - \phi$$

$$KE_{max} = \frac{6.63 \times 10^{-34} \times 3.0 \times 10^8}{3.6 \times 10^{-7}} - (2.5 \times 1.6 \times 10^{-19})$$

$$KE_{max} = 1.53 \times 10^{-19}\,\text{J} \quad (\approx 0.96\,\text{eV})$$

examiner's tip

For consistent working, it is important to convert the energy from electronvolts to joules.

(9) (a) A line of best-fit is drawn with intercept of 4.8×10^{14} Hz with the f-axis.

(b) Photoelectrons are **not** released from the metal surface by the incident radiation.

(c) $f_0 \approx 4.8 \times 10^{14}$ Hz

examiner's tip	At the threshold frequency, the photoelectrons are just released from the metal surface by the incoming photons. The freed electrons have no kinetic energy. The intercept of the line with the frequency axis is therefore the value of the threshold frequency.

(d) $hf_0 = \phi$
$\phi \approx 6.63 \times 10^{-34} \times 4.8 \times 10^{14}$
$\phi \approx 3.2 \times 10^{-19}$ J

(e) $hf = \phi + E_k$ (Einstein's photoelectric equation)
$\therefore E_k = \mathbf{h}f - \phi$
Comparing this with $y = \mathbf{m}x + c$, the equation for a straight line, the gradient must be h, the Planck constant.

 **Examining Group**

Physics

Time: 1 hour   Maximum marks: 60

Instructions

Answer **all** questions in the spaces provided. Show all steps in your working.

The marks allocated for each question are shown in brackets.

Any data required for a question are given where appropriate.

Grading

Boundary for A grade   48/60

Boundary for C grade   36/60

---

1   The drag force F acting on a car travelling at a constant speed v is given by the equation

$$F = k\,v^2$$

where k is a constant for a given car.

(a) Determine the unit for k and suggest one factor related to the car that may affect the magnitude of k.

....

....

.... [2]

(b) For the car travelling at a constant velocity of 27 m s^{-1}, the drag force is 670 N.

     (i) State and explain the magnitude of the motive force produced by the car engine.

     ....

     .... [2]

     (ii) Calculate the value for k.

     ....

     .... [2]

(iii) Calculate the power developed by the engine when the car is travelling at a constant speed of 27 m s^{-1}.

....

.... [2]

(iv) Show that the motive power P developed by the engine is given by

$$P \propto v^3$$

Hence determine the change in the motive power of the engine when the speed of the car is **reduced** by 20%. Suggest one advantage of reducing car speed.

....

....

....

.... [3]

[Total: 11]

2 The graph below shows the current–voltage characteristic of a lamp provided by a manufacturer.

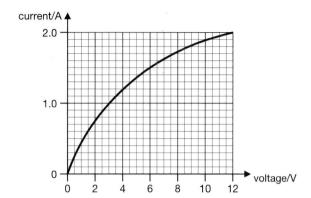

(a) State how the resistance of the lamp changes as the voltage across it is increased.

....

.... [1]

(b) A student buys two identical lamps from the manufacturer. The lamps are connected to a 12 V d.c. supply that has negligible internal resistance. Determine the total resistance of the circuit when both lamps are connected to the supply

   (i) in a series combination,

   ....

   .... [2]

  (ii) in a parallel combination.

   ....

   .... [2]

(c) The diagram shows **one** of the lamps and a 10 Ω resistor connected in series to a battery.

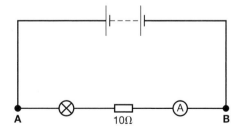

The current measured by the ammeter is 1.0 A. The ammeter has negligible resistance. Calculate the potential difference across **A B**.

....

....

.... [5]

[Total: 10]

3 A metal wire of cross-sectional area 3.2×10^{-8} m² has a length 2.5 m. At room temperature, the wire has a resistance of 4.6 Ω.

(a) Calculate the resistivity of the material the wire is made from.

....

.... **[3]**

(b) The diagram shows the wire wound into a tight coil and connected to a supply of negligible internal resistance.

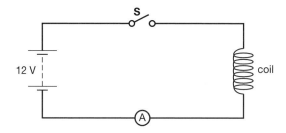

When the switch **S** is closed, the coil eventually gets 'red hot'. The graph below shows the variation of the circuit current *I* with time *t*.

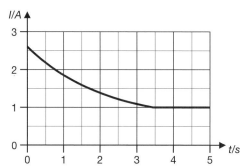

(i) Show why the current at $t = 0$ s is 2.6 A. Explain your answer.

....

.... **[2]**

(ii) With reference to the motion of the electrons within the material of the wire, explain the shape of the graph.

....

....

....

.... **[3]**

[Total: 8]

4 (a) State two conditions necessary for producing a stationary wave.

....

....

.... [2]

(b) A microwave transmitter is placed in front of a metal sheet. A standing wave is set up between the metal plate and the transmitter. The microwaves have a wavelength of 2.8 cm.

(i) Calculate the separation between two adjacent antinodes.

....

....

.... [2]

(ii) Draw a diagram to show the standing wave pattern between the metal plate and the loudspeaker.

[2]

(iii) The microwave transmitter is now replaced by one emitting waves of higher frequency. The position of the transmitter is adjusted so that another standing wave is produced between it and the metal plate. State and explain the change, if any, to the standing wave pattern produced.

....

....

....

.... [2]

[Total: 8]

5 **(a)** State three assumptions of the kinetic theory of gases.

....

....

.... [3]

(b) State the meaning of the symbols in the equation

$$pV = \tfrac{1}{3} Nm\langle c^2 \rangle.$$

....

....

....

....

.... [5]

(c) Air consists principally of oxygen and nitrogen. Both gases may be considered to behave as ideal gases at room temperature of 20°C.

(i) Show that the mean translational kinetic energy E_k of each molecule of an ideal gas is given by

$$E_k = \tfrac{3}{2} kT$$

where k is the Boltzmann constant.

....

....

.... [3]

(ii) Calculate the mean translational kinetic energy E_k of a nitrogen molecule at 20°C.
Data: $k = 1.38 \times 10^{-23}\,\text{J K}^{-1}$

....

.... [2]

(iii) Oxygen molecules are more massive than nitrogen molecules.
State and explain how the answer to **(c)(ii)** would change, if at all, for an oxygen molecule.

....

.... [2]

(iv) Explain what is meant by the **internal energy** of a gas. Calculate the total internal energy of one mole of nitrogen gas.
Data: $N_A = 6.02 \times 10^{23}\,mol^{-1}$

....

....

....

....

.... [3]

[Total: 18]

6 (a) State the speed of a photon in free space.

.... [1]

(b) A light-emitting diode (LED) emits yellow light of wavelength 550 nm at a rate of 30 mW.
Data: $h = 6.63 \times 10^{-34}\,J\,s$
$c = 3.0 \times 10^8\ m\,s^{-1}$

(i) Calculate the energy of a single photon of yellow light.

....

....

.... [2]

(ii) Calculate the number of photons emitted by the light-emitting diode per second.

....

....

.... [2]

[Total: 5]

 **Examining Group**

Physics

Time: 1 hour   Maximum marks: 60

Instructions

Answer **all** questions in the spaces provided. Show all steps in your working.
The marks allocated for each question are shown in brackets.
Any data required for a question are given where appropriate.

Grading
Boundary for A grade   48/60
Boundary for C grade   36/60

---

1 (a) Define tensile stress.

....

.... **[2]**

(b) The diagram below shows a display shelf in a shop.

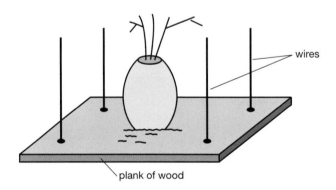

wires

plank of wood

The shelf consists of a plank of wood that is supported by four vertical wires. The mass of the plank of wood is 1.4 kg. An object of mass 3.5 kg is placed half way along the length of the shelf. Each supporting wire has a diameter of 1.6 mm and a length of 90 cm. The material of the wire has a Young modulus of 1.8×10^{10} Pa.

Data: $g = 9.8\,\text{N kg}^{-1}$

For each supporting wire, calculate

(i) the tension,

....

....

.... [2]

(ii) the extension,

....

....

.... [3]

(iii) the elastic strain energy.

....

....

.... [2]

(c) State one assumption made when doing the calculations in **(b)(ii)** and **(b)(iii)**.

....

.... [1]

[Total: 10]

2 A car of mass 820 kg is travelling on a level road at a velocity of 25 m s^{-1}. The driver sees an obstacle ahead and applies her brakes. Her reaction time is 0.60 s and the total distance travelled by the car before it stops is 80 m.

(a) Calculate the kinetic energy of the car when travelling at 25 m s^{-1}.

....

.... [2]

(b) Determine the distance travelled by the car during the time taken by the driver to react.

....

.... [1]

(c) Determine the braking distance of the car.

....

.... [1]

(d) Calculate the braking force acting on the car.

....

....

.... [3]

(e) Explain how your answer to **(d)** would change when the braking distance remains constant, but

   (i) the car was travelling up a hill as the brakes were applied,

   ....

   .... [1]

   (ii) the initial speed of the car was 50 m s^{-1}.

   ....

   .... [1]

[Total: 9]

3 In school and college laboratories, americium-241 is used as a source of alpha-particles. A particular americium source has an activity of 1.2×10^5 Bq and is housed in a lead container of mass 80 g. A nucleus of americium ($^{241}_{95}$Am) decays into neptunium (Np). Each alpha-particle has a kinetic energy of 5.4 MeV.

Data: $1\,eV = 1.6 \times 10^{-19}$ J

(a) Define *activity* of a source.

....

.... [1]

(b) Write down the nuclear decay equation for the nucleus of americium-241.

....

.... [2]

(c) Calculate the energy released per second by the americium source.

....

....

.... [3]

(d) The alpha-particles are readily absorbed by the lead container. Estimate the change in the temperature of the lead container over a period of 1.0 years assuming that all the kinetic energy of the alpha-particles is absorbed and contributes to the heating of the lead container. (You may assume that there is no heat lost to the surroundings.)

Data: specific heat capacity of lead is 130 J kg K^{-1}.

....

....

....

.... [3]

[Total: 9]

4 (a) Define the electromotive force (e.m.f.) of a cell.

....

....

.... [2]

(b) State the main difference between the definition for potential difference and electromotive force.

.... [1]

(c) The diagram below shows an electrical circuit.

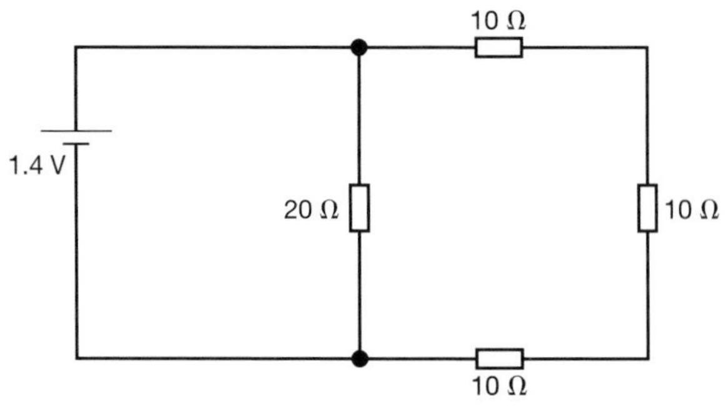

The cell has negligible internal resistance.

(i) Calculate the total resistance of the circuit.

....

....

.... [2]

(ii) Calculate the amount of chemical energy transformed by the cell in a time of 5.0 minutes.

....

....

.... [2]

(d) The diagram shows an electrical circuit containing a variable resistor and a thermistor.

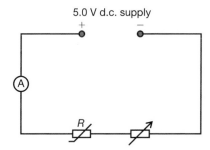

The d.c. supply has negligible internal resistance. As the slider of the variable resistor is moved from one extreme to the other, the current in the circuit changes from 25 mA to 200 mA. The temperature of the thermister remains constant.
Determine

  (i) the resistance R of the thermistor,

....

....

.... [2]

  (ii) the maximum resistance of the variable resistor.

....

....

.... [2]

[Total: 11]

5 (a) A ray of light travelling in air enters a glass block. State the effect on

   (i) the speed of light,

   .... [1]

   (ii) the frequency of light,

   .... [1]

   (iii) the wavelength of light.

   .... [1]

(b) The diagram shows two short pulses of light, one travelling in a glass rod and the other in vacuum.

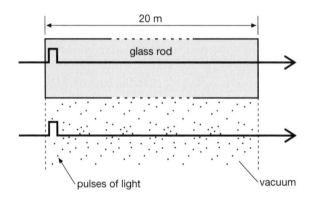

In a certain time interval, the pulse travelling in vacuum covers a distance of 20 m. Calculate the distance travelled by the pulse in the glass rod in the same interval.
Data: refractive index of glass = 1.48
$c = 3.00 \times 10^8 \, \text{m s}^{-1}$

....

....

....

.... [2]

(c) Red light of wavelength 6.8×10^{-7} m is incident normally on a metal plate containing two parallel slits separated by 0.1 mm. An interference pattern is formed on a screen placed 2.5 m away from the slits.

(i) Draw a diagram to illustrate the appearance of the interference pattern formed on the screen.

[2]

(ii) Calculate the separation between two neighbouring bright patches (fringes) of light in the pattern.

....

....

....

.... [3]

(iii) Describe the effect on the pattern shown in **(c)(i)** when one of the slits is covered up so that light only passes through one of the slits.

....

....

.... [2]

[Total: 12]

6 (a) The diagram shows two energy levels for a particular type of atom.

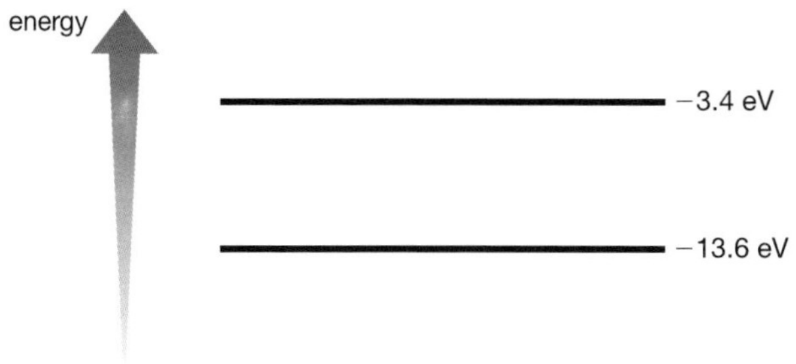

energy

————————————————— −3.4 eV

————————————————— −13.6 eV

   (i) Explain why energy levels have negative values.

....

.... [1]

   (ii) On the diagram indicate with an arrow the transition made by an electron responsible for emitting a photon from the atom. [1]

(b) A laser produces a monochromatic light of wavelength 6.4×10^{-7} m.
The output light power of the laser is 6.0 mW.

   (i) Calculate the energy of a single photon of light from the laser.
   Data: $h = 6.63 \times 10^{-34}$ J s
   $c = 3.0 \times 10^{8}$ m s^{-1}

....

....

.... [3]

   (ii) Hence determine the number of photons released from the laser per second.

....

....

.... [2]

(c) Without any further calculations, state and explain the effect on the number of photons released per second if the 6.0 mW laser in **(b)** were to produce high frequency X-rays.

....

....

.... [2]

[Total: 9]

 **Examining Group**

Physics

Time: 1 hour   Maximum marks: 60

Instructions

Answer **all** questions in the spaces provided. Show all steps in your working.

The marks allocated for each question are shown in brackets.

Any data required for a question are given where appropriate.

Grading

Boundary for A grade   48/60

Boundary for C grade   36/60

---

1   A boat is travelling at a constant velocity of 12 m s^{-1} due North. The wind is blowing at a constant velocity of 5.0 m s^{-1} from the west.

(a) Velocity is a **vector** quantity.

Explain what is meant by a vector and give an example of another vector quantity.

....

.... **[2]**

(b) With the aid of a vector diagram, calculate the total velocity of the boat.

....

....

.... **[4]**

(c) Determine the northerly displacement of the boat in a time of 10 minutes.

....

.... **[2]**

[Total: 8]

2 (a) Define moment of a force.

....

.... [2]

(b) The diagram shows a section of a bridge that is resting on two concrete supports.

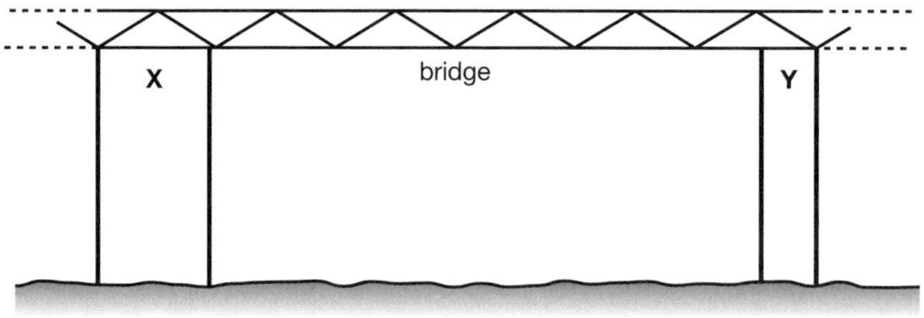

The support **X** has a cross-sectional area of 8.0 m² and the other support, **Y**, has a cross-sectional area of 6.0 m². The mass of the uniform horizontal section of the bridge between the two supports is 7.5×10^5 kg.
Data: $g = 9.8\,\text{N kg}^{-1}$

(i) Determine the force exerted at each support.

....

....

....

.... [2]

(ii) State and explain which of the two supports experiences the greatest stress.

....

....

....

.... [2]

(iii) The diagram below shows a railway engine of mass 8.2×10^4 kg crossing the bridge.

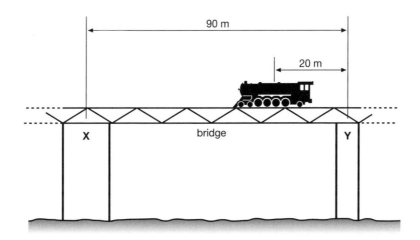

The engine is 20 m from the support **Y**. Calculate the magnitude of the force *F* exerted by the support **Y** on the bridge.

....

....

....

....

.... **[4]**

[Total: 10]

3 (a) The diagram shows a circuit designed to monitor the temperature of a small greenhouse.

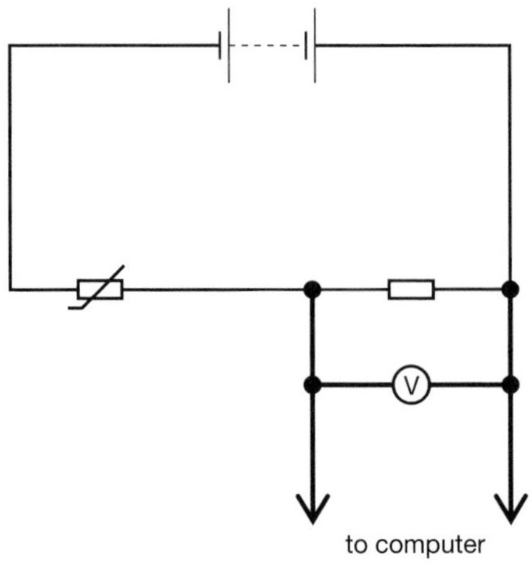

to computer

The d.c. supply has negligible internal resistance and the voltmeter has an infinite resistance. Describe what happens to the reading on the voltmeter as the temperature of the negative temperature coefficient (NTC) thermistor is decreased.

....

....

.... [3]

(b) A cell of e.m.f. 1.50 V has an internal resistance of 0.40 Ω. The cell is connected to a variable resistor.

    (i) The variable resistor is set to 0.10 Ω. Show that

       1. the current in the circuit is 3.0 A,

       ....

       .... [2]

       2. the power dissipated by the variable resistor is 0.9 W.

       ....

       .... [2]

    (ii) Complete the table below to show how the current I in the circuit and power P dissipated by the external variable resistor are affected by its resistance R.

R/Ω	I/A	P/W
0	5.00	0
0.1	3.00	0.90
0.4		
0.7		
1.0		

[3]

    (iii) Use your answer to **(b)(ii)** to suggest when the power dissipated by the external load (variable resistor) is a maximum.

    .... [1]

[Total: 11]

4 (a) A small amount of ice at a temperature of $-10°C$ is gradually heated until it first changes to water and then to steam. Explain, in molecular terms, what happens to the energy supplied to the ice as it first changes to water and then into steam.

....

....

....

....

....

....

.... [4]

(b) Suggest why steam scalds are more serious than scalds by water at the same temperature of 100°C.

....

....

.... [2]

(c) A small heater is used to heat a liquid in a well-lagged container. The mass of the liquid is 120 g. The heater is connected to a d.c. supply of negligible internal resistance and having an e.m.f. of 12 V. The current drawn from the supply is measured as 4.0 A. In a time of 2.5 minutes, the temperature of the liquid increases from 21°C to 35°C. Determine the specific heat capacity of the liquid.

....

....

....

....

....

.... [4]

[Total: 10]

5 (a) Define the frequency of a wave.

.... [1]

(b) Show that the speed v of a wave is given by the equation

$$v = f\lambda$$

where f is the frequency of the wave and λ is the wavelength.

....

....

....

.... [3]

(c) A long rope is oscillated at one end. The rope is oscillated at a frequency of 5.0 Hz. The speed of the transverse wave on the rope is 1.6 m s⁻¹. The diagram shows the profile of the rope at a particular time.

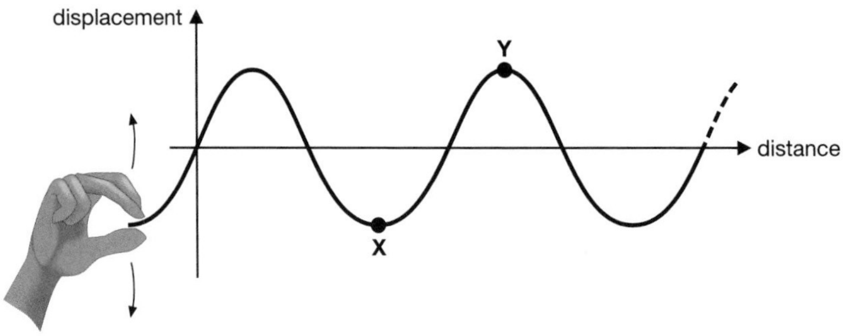

(i) State the phase difference between the points **X** and **Y**.

.... [1]

(ii) Explain how the profile of the rope would change after a time of 0.2 s.

....

....

.... [2]

(iii) Calculate the wavelength of the wave.

....

....

.... [2]

[Total: 9]

6 A clean plate of zinc is charged negatively. When ultraviolet light is incident on the plate, photoelectric emission may take place.

(a) Explain what is meant by photoelectric emission and describe what happens to the charge on the zinc plate.

....

....

....

.... [3]

(b) The surface of caesium is exposed to all the visible wavelengths that lie in the range 4.0×10^{-7} m to 7.0×10^{-7} m. The work function of caesium is 1.4 eV.

Data: $h = 6.6 \times 10^{-34}$ J s

$c = 3.0 \times 10^{8}$ m s^{-1}

$e = 1.6 \times 10^{-19}$ C

mass of electron $= 9.1 \times 10^{-31}$ kg

(i) Calculate the maximum kinetic energy of a photoelectron.

....

....

.... [3]

(ii) Calculate the maximum speed of a photoelectron emitted from the surface of caesium.

....

....

.... [3]

(iii) Calculate the de Broglie wavelength of a photoelectron emitted in **(b)(i)**.

....

....

.... [3]

[Total: 12]

(1) (a) $k = \dfrac{F}{v^2}$

Therefore k has units $N\,s^2\,m^{-2}$.
The drag would depend on the frontal area of the car.

(b) (i) The car is travelling at constant velocity and therefore has no acceleration.
According to $F = ma$, the **net** force on the car will be zero.
Hence, the force provided by the engine, the motive force = 670 N.

(ii) $k = \dfrac{F}{v^2}$

$k = \dfrac{670}{27^2}$

$k = 9.19 \times 10^{-1} \approx 9.2 \times 10^{-1} \ N\,s^2\,m^{-2}$

(iii) Power $= Fv$
$P = 670 \times 27$
$P = 1.81 \times 10^4 \approx 1.8 \times 10^4$ W

(iv) $P = Fv$
$P = (kv^2)v = kv^3$
k is a constant, therefore $P \propto v^3$

With a 20% reduction in speed, the power will decrease to $(0.80)^3$ of its original value. The engine power will therefore be 51% of its previous value.
(There is change in power of 49%.)

One possible advantage would be using less fuel and therefore reduced environmental pollution.

examiner's tip	You can do the question by calculating the drag on the car at the new speed using $F = kv^2$ and then applying $P = Fv$. The route shown above is concise and shows that the actual value of k does not matter in this case. It is also advantageous to use the information provided by the examiners. In this case, the relationship $P \propto v^3$.

(2) (a) As the p.d. increases, the resistance of the lamp increases.

examiner's tip	It is very sad when candidates guess the answer here. You can play safe and calculate the value of the resistance at low and high voltages using the graph.

(b) (i) The bulbs are identical, therefore the p.d. across each is 6.0 V
From the graph, current = 1.5 A

$R = \dfrac{V}{I} = \dfrac{6.0}{1.5} = 4.0\,\Omega$

Total resistance, $R_T = 4.0 + 4.0 = 8.0\,\Omega$

(ii) Since the bulbs are connected in parallel, the p.d. across each lamp is 12 V.
From the graph, current = 2.0 A

$$R = \frac{V}{I} = \frac{12}{2.0} = 6.0 \, \Omega$$

$$R_T = \frac{R_1 R_2}{(R_1 + R_2)}$$

$$R_T = \frac{6.0 \times 6.0}{12.0}$$

$$R_T = 3.0 \, \Omega$$

examiner's tip   This question cannot be done without the graph given at the start of the question.
Use all the information given by the examiners.

(c) From the graph, when current is 1.0 A the potential difference across the bulb is 3.0 V.

$$R_{BULB} = \frac{V}{I} = \frac{3.0}{1.0} = 3.0 \, \Omega$$

$$R_T = R_1 + R_2 = 10 + 3.0$$
$$R_T = 13 \, \Omega$$

$$V = IR$$
$$V = 1.0 \times 13$$
$$V = 13 \, V$$

(3) (a) $R = \dfrac{\rho \ell}{A}$

$$\rho = \frac{RA}{\ell} = \frac{4.6 \times 3.2 \times 10^{-8}}{2.5}$$

$$\rho = 5.89 \times 10^{-8} \approx 5.9 \times 10^{-8} \, \Omega \, m$$

(b) (i) When the switch is closed, the coil is at room temperature and therefore has resistance of 4.6 Ω.

$$I = \frac{V}{R} = \frac{12}{4.6} = 2.61 \, A$$

(ii) The wire starts to heat up as current passes.
This leads to greater vibration of the ions and the electrons collide more frequently with the ions.
The resistance of the wire therefore increases, which is shown by the decreasing current.
Eventually, the coil attains a constant temperature, therefore the current is constant.

examiner's tip   The examiners are looking for at least three distinct comments from the candidate.
It is important that the candidate has something to write about the motion of the
electrons within the wire.

(4) (a) A standing wave arises when two waves having the same wavelength superimpose as they travel in opposite directions.

(b) **(i)** The separation between two adjacent antinodes $= \dfrac{\lambda}{2} = \dfrac{2.8}{2} = 1.4$ cm

(ii)

examiner's tip	It is important to draw a clear diagram that is well-annotated. A decent sketch should show that the separation between adjacent antinodes is equal to $\dfrac{\lambda}{2}$.

(iii) Higher frequency implies shorter wavelength of the microwaves.
Hence the separation between adjacent antinodes (or nodes) will decrease.

(5) (a) Any *three* from:
The molecules have a negligible volume compared with the volume of the container.
The molecules collide elastically with each other (or the container walls).
The forces between the molecules are negligible, except during collisions.
The motion of the molecules is random.
The time for molecular collisions is negligible compared to the time between the collisions.
There are a very large number of molecules inside the container.

(b) p is the pressure exerted by the gas.
V is the volume of the gas (or the container).
N is the number of molecules.
m is the mass of each gas molecule.
$\langle c^2 \rangle$ is the mean square speed of the molecules.

(c) **(i)** $pV = nRT$
$nRT = \frac{1}{3}Nm\langle c^2 \rangle$
For one mole of gas, $n = 1$ and $N = N_A$, the Avogadro constant.
$RT = \frac{1}{3}N_A m \langle c^2 \rangle$

$$\frac{1}{3}m\langle c^2 \rangle = \left(\frac{R}{N_A}\right)T$$

But $k = \dfrac{R}{N_A}$

$\therefore \frac{1}{2}m\langle c^2 \rangle = \frac{3}{2}kT$
$\frac{1}{2}m\langle c^2 \rangle$ is the mean translational kinetic energy, E_k of the molecule.

(ii) $E_k = \frac{3}{2} \times 1.38 \times 10^{-23} \times (273 + 20)$
$E_k = 6.07 \times 10^{-21} \approx 6.1 \times 10^{-21}$ J

examiner's tip	It is vital that the temperature is written in kelvin. T is the thermodynamic temperature.

(iii) The mean translational kinetic energy for the oxygen molecule will be the same. This is because $E_k \propto T$, hence independent of the mass.

examiner's tip

At the same temperature T, the mean translational kinetic energy E_k for all types of molecules is the same. This is because $E_k \propto T$. The more massive particles will move slowly, however,

$$\tfrac{1}{2} m \langle c^2 \rangle$$

will be the **same** for all molecules.

(iv) Internal energy = sum of K.E. and P.E. of molecules.
E = energy of one mole of gas
$E = 6.07 \times 10^{-21} \times 6.02 \times 10^{23}$ (Assume P.E. = 0)
$E = 3.65 \times 10^3 \approx 3.7 \times 10^3 \, \text{J}$

(6) (a) The speed of a photon in free space is c, $3.0 \times 10^8 \, \text{m s}^{-1}$

examiner's tip

The speed of a photon is the same as the speed of electromagnetic waves in a vacuum.

(b) (i) $E = hf = \dfrac{hc}{\lambda}$

$E = \dfrac{6.63 \times 10^{-34} \times 3.0 \times 10^8}{550 \times 10^{-9}} = 3.62 \times 10^{-19} \, \text{J}$

(ii) number of photons released per second = power / energy of each photon

$\text{number} = \dfrac{0.030}{3.62 \times 10^{-19}} = 8.3 \times 10^{16} \, \text{s}^{-1}$

(1) (a) tensile stress $= \dfrac{\text{force}}{\text{cross-sectional area}}$

(b) (i) tension $= \dfrac{\text{total weight}}{4}$

tension $= \dfrac{mg}{4} = \dfrac{4.9 \times 9.8}{4}$ (m = 1.4 + 3.5 = 4.9 kg)

tension $= 12\,\text{N}$

examiner's tip	Do not forget that the weight is shared equally by the four wires. Hence the tension in each wire is a quarter of the total weight of the plank and the object.

(ii) Young modulus = stress/strain

$$E = \dfrac{F/A}{x/L}$$

$$x = \dfrac{FL}{EA} = \dfrac{12 \times 0.9}{1.8 \times 10^{10} \times (\pi \times 0.0008^2)}$$

$$x = 2.98 \times 10^{-4}\,\text{m} \approx 0.30\,\text{mm}$$

examiner's tip	You can calculate the extension by first determining the stress in each wire, using stress $= \dfrac{F}{A}$. Then the strain using strain $= \dfrac{\text{stress}}{E}$ and then finally, the extension using extension = strain × L.

(iii) strain energy $= \frac{1}{2}Fx$

strain energy $= \frac{1}{2} \times 12 \times 2.98 \times 10^{-4}$

strain energy $= 1.8\,\text{mJ}$

(c) The elastic limit has not been exceeded **or** Hooke's law is obeyed by the wires.

(2) (a) $KE = \frac{1}{2}mv^2$

$KE = \frac{1}{2} \times 820 \times 25^2$

$KE = 2.56 \times 10^5 \, J \approx 2.6 \times 10^5 \, J$

(b) distance $= 25 \times 0.60 = 15 \, m$

(c) braking distance $= 80 - 15 = 65 \, m$

(d) work done by braking force $= KE$ of car

$F \times 65 = 2.56 \times 10^5$ (*F* is the magnitude of the braking force)

$$F = \frac{2.56 \times 10^5}{65} = 3.94 \times 10^3 \, N$$

$F \approx 3.9 \, kN$

---

examiner's tip

You can use the equation of motion
$$v^2 = u^2 + 2as$$
and $F = ma$ to get the answer. If you do this, the braking force F will be
$$F = \frac{-mu^2}{2s}.$$
Also, do not forget the correct distance is 65 m and not 80 m or even 15 m.

(e) (i) The braking force would be less as the car transfers some of its kinetic energy into gravitational potential energy.

---

examiner's tip

This is a tough question. By conservation of energy we have

work done by braking force $= KE$ of car $-$ gain in gravitational PE

Therefore $Fs = KE$ of car $-$ gain in gravitational PE

$Fs < KE$ of car

Hence for the same braking distance s, the braking force F would be less than before.

(ii) For a constant braking distance, the braking force is directly proportional to the square of the initial speed (see 'examiner's tip' above). Hence the braking force will have to increase by a factor of four if the distance is to be the same.

(3) (a) The activity of a source is the rate of decay of the nuclei with respect to time.

> **examiner's tip**
>
> **You may also define activity as the *'number of disintegrations of the nuclei per unit time'*. It is important that you make reference to nuclei rather than atoms when defining activity.**

(b) $^{241}_{95}Am \rightarrow {}^{4}_{2}He + {}^{237}_{93}Np$

(c) energy released per second = activity × kinetic energy of each α-particle
energy released per second = $1.2 \times 10^{5} \times (5.4 \times 10^{6} \times 1.6 \times 10^{-19})$
energy released per second = $1.04 \times 10^{-7}\,J\,s^{-1}$

(d) $E = mc\Delta T$

$E = 1.04 \times 10^{-7} \times 365 \times 24 \times 3600$

$E = 3.28\,J$

$$\Delta T = \frac{E}{mc}$$

$$\Delta T = \frac{3.28}{0.080 \times 130}$$

$\Delta T \approx 0.32°C$

(4) (a) electromotive force = energy transformed to **electrical** per unit charge.

(b) Both e.m.f. and p.d. are defined as energy per unit charge, but for e.m.f. the energy is transformed from some other form, e.g. chemical to electrical, whereas for p.d. the energy is transformed from electrical to some other form.

(c) **(i)** Total resistance of the three $10\,\Omega$ in series combination $= 30\,\Omega$
Total resistance of circuit $= R_T$, with

$$\frac{1}{R_T} = \frac{1}{30} + \frac{1}{20}$$

$$R_T = \frac{20 \times 30}{20 + 30} = 12\,\Omega$$

(ii) energy = power $\times$ time

$$\text{energy} = \left(\frac{V^2}{R}\right) \times \text{time}$$

$$\text{energy} = \frac{1.4^2}{12} \times (5.0 \times 60) = 49\,\text{J}$$

> **examiner's tip** You can calculate the total energy dissipated by each resistor but the answer above is quicker because it uses the total resistance of the circuit and the e.m.f. of the cell.

(d) **(i)** $R = \dfrac{V}{I} = \dfrac{5.0}{0.200}$

$R = 25\,\Omega$

(ii) $R = \dfrac{V}{I} = \dfrac{5.0}{0.025}$

$R = 200\,\Omega$
The maximum resistance R_{max} of the variable resistor is
$R_{max} = 200 - 25 = 175\,\Omega$

> **examiner's tip** The diagram below illustrates the circuit for the two extreme values of the current.
>
>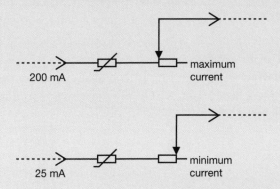
>
> Remember for a given e.m.f., the current is inversely proportional to the resistance of the circuit.

(5) (a) (i) The speed decreases.

(ii) The frequency remains the same.

(iii) The wavelength decreases.

(b) Time taken for pulse in vacuum to travel a distance of 20 m $= \dfrac{20}{3.00 \times 10^8}$

$$= 6.67 \times 10^{-8}\,\text{s}$$

Speed of pulse in glass $= \dfrac{3.00 \times 10^8}{1.48} = 2.027 \times 10^8\,\text{m s}^{-1}$

Distance travelled by pulse in glass rod $= 2.027 \times 10^8 \times 6.67 \times 10^{-8} = 13.5\,\text{m}$

(c) (i)

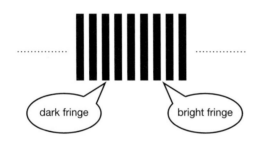

dark fringe

bright fringe

(ii) $\lambda = \dfrac{ax}{D}$

$$x = \frac{\lambda D}{a} = \frac{6.8 \times 10^{-7} \times 2.5}{0.0001} = 0.017\,\text{m}$$

$$x = 1.7\,\text{cm}$$

(iii) The interference pattern disappears (no bright and dark fringes). Instead, the light from one of the slits gives a diffraction pattern.

(6) (a) (i) Negatively charged electrons are bound to the positive nucleus.
External energy is required to either excite the atom or to remove the electron completely from the influence of the positive nucleus.

(ii) Transition shown from the $-3.4\,eV$ energy level to the $-13.6\,eV$ energy level.

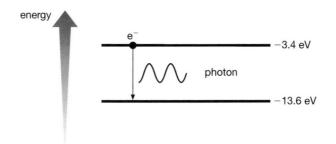

examiner's tip An electron making the above transition is losing energy. In terms of energy, the atom is much more stable after this event. Since energy must be conserved, the energy of the electron is transformed into a **photon** of electromagnetic radiation.

(b) (i) $f = \dfrac{c}{\lambda}$

$$f = \frac{3.0 \times 10^8}{6.4 \times 10^{-7}}$$

$f = 4.69 \times 10^{14}\,Hz$

$E = hf$

$E = 6.63 \times 10^{-34} \times 4.69 \times 10^{14}$

$E = 3.11 \times 10^{-19} \approx 3.1 \times 10^{-19}\,J$

(ii) photons release per second = power of laser / energy of each photon

$$N = \frac{6.0 \times 10^{-3}}{3.11 \times 10^{-19}}$$

$N = 1.93 \times 10^{16} \approx 1.9 \times 10^{16}\,s^{-1}$

(c) The energy of a single X-ray photon is greater.
Hence for the **same** power, there are fewer photons released in a given time.

(1) (a) A vector quantity has both magnitude and direction.
Any one from:
momentum, force and acceleration.

(b)

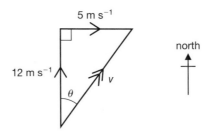

The resultant velocity = v
$v = \sqrt{12^2 + 5.0^2}$
$v = 13\,\mathrm{m\,s^{-1}}$

$\theta = \tan^{-1}\left(\dfrac{5.0}{12}\right) \approx 23°$

examiner's tip	The answer requires both magnitude and direction of the resultant velocity. You must therefore determine the angle shown. The boat travels at a bearing of 023°.

(c) distance = speed × time
distance = $12 \times (10 \times 60) = 7.2\,\mathrm{km}$

examiner's tip	The distance travelled in the northern direction requires the velocity in this direction. The correct velocity is $12\,\mathrm{m\,s^{-1}}$ and not any other value.

(2) (a) Moment of force = force × **perpendicular** distance of the line of action of the force from a point or pivot.

examiner's tip	You would lose a mark if no reference is made to the distance being 'perpendicular'. If you have problems with definitions, then try to support your answer with a labelled diagram.

(b) (i) weight of bridge = mg

force at each support $= \dfrac{mg}{2} = \dfrac{7.5 \times 10^5 \times 9.8}{2}$

force at each support $= 3.7 \times 10^6\,\mathrm{N}$

(ii) stress $= \dfrac{\text{force}}{\text{cross-sectional area}}$

The force on each support is the same, therefore
stress $\propto$ 1/cross-sectional area
The support **Y** experiences the greatest stress.

(iii) clockwise moments about **X** = anticlockwise moments about **X**

$$(70 \times 8.2 \times 10^4 \times 9.8) + (45 \times 7.5 \times 10^5 \times 9.8) = F \times 90$$

$$F = \frac{3.87 \times 10^8}{90} = 4.3 \times 10^6 \, \text{N}$$

examiner's tip Do not forget to include the weight of the bridge in this calculation and also that the weight of the horizontal section of the bridge acts at its midpoint.

(3) (a) Decreasing the temperature of the thermistor increases its resistance.
The potential difference across the thermistor increases, therefore the potential difference across the resistor decreases.

examiner's tip Remember the circuit is a potential divider circuit.

(b) (i) 1. $I = \dfrac{V}{R}$

$$I = \frac{1.5}{(0.40 + 0.10)} = 3.00 \, \text{A}$$

2. $P = I^2 R$

$$P = 3.00^2 \times 0.10 = 0.90 \, \text{W}$$

(ii)

R/Ω	I/A	P/W	
0	5.00	0	
0.1	3.00	0.90	
0.4	**1.88**	**1.41**	Maximum power
0.7	**1.36**	**1.30**	
1.0	**1.07**	**1.15**	

(iii) The power dissipated in the external resistor is a maximum when its resistance is equal to the resistance of the internal resistor.

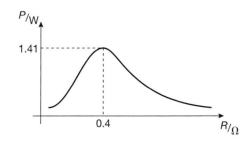

(4) (a) As the temperature of the solid ice increases from −10°C to 0°C, the vibrational energy of the molecules increases and therefore the internal energy of the ice increases.

At 0°C, the energy absorbed is used by the ice to change state (or phase) from a solid to a liquid. Molecular bonds are broken but the temperature stays the same. The energy absorbed is known as the latent heat of fusion. Once all the bonds are broken, further absorption of energy increases the kinetic energy of the water molecules.

This kinetic energy increases as the temperature of the water rises until it reaches 100 °C. At this temperature, energy is absorbed to change the phase from water to steam. The energy absorbed by the water to change phase to steam is referred to as the latent heat of vaporisation. The temperature remains constant at 100°C as all the bonds are broken and all the water changes to steam.

(b) Water will cool down to the body temperature and the energy is conducted through the skin. This energy is much smaller than the energy released by the steam as it also changes phase from steam to water.

(c) heat supplied by heater = heat absorbed by the water

$Vlt = mc\Delta T$

$$c = \frac{Vlt}{m\Delta T} = \frac{12 \times 4.0 \times (2.5 \times 60)}{0.120 \times (35 - 21)}$$

$c = 4300 \, \text{J} \, \text{kg}^{-1} \, \text{K}^{-1}$

(5) (a) Frequency is the number of waves passing a point per unit time.

(b) speed = distance / time
In a time equal to the period T of the wave, the wave will progress a distance equal to the wavelength λ. Therefore the speed v of the wave is given by

$$v = \frac{\lambda}{T}.$$

However, $\frac{1}{T}$ is equal to the frequency f of the wave. Hence

$$v = \frac{1}{T} \times \lambda = f\lambda.$$

examiner's tip   **You must show all the steps of this proof and try to explain each step.**

(c) (i) The points **X** and **Y** are separated by half a wavelength, therefore the phase difference will be 180° or π radians.

(ii) The period of the wave is 0.2 s. Hence there is no change to the profile of the rope because in a time equal to the period of the wave, the wave progresses by a distance equal to one whole wavelength.

examiner's tip   **You must appreciate the link between frequency and period. You cannot get anywhere in this question if you have not spotted that $T = \frac{1}{f}$.**

(iii) $v = f\lambda$

$$\lambda = \frac{v}{f}$$

$$\lambda = \frac{1.6}{5.0}$$

$$\lambda = 0.32 \text{ m}$$

(6) (a) Photoelectric emission is when electrons are emitted from the surface of a metal that is exposed to electromagnetic radiation.

The zinc plate loses electrons. Since electrons carry negative charge, the magnitude of the charge on the plate decreases.

(b) (i) $hf = \phi + KE_{max}$

$$KE_{max} = \frac{hc}{\lambda} - \phi = \frac{6.6 \times 10^{-34} \times 3.0 \times 10^8}{4.0 \times 10^{-7}} - (1.4 \times 1.6 \times 10^{-19})$$

$$KE_{max} = 2.71 \times 10^{-19}\,J \approx 2.7 \times 10^{-19}\,J$$

examiner's tip	The maximum kinetic energy of the electron will correspond to the **shortest** wavelength of visible light. Most likely there will be a mark reserved for the use of the correct wavelength. You must also convert the electronvolts into joules.

(ii) $\frac{1}{2}mv^2 = 2.71 \times 10^{-19}$

$$v = \sqrt{\frac{2 \times 2.71 \times 10^{-19}}{9.1 \times 10^{-31}}}$$

$$v = 7.7 \times 10^5\,m\,s^{-1}$$

examiner's tip	A common error in examinations is calculating v^2 rather than v. Do not forget to square root your answer.

(iii) $\lambda = \frac{h}{mv}$

$$\lambda = \frac{6.6 \times 10^{-34}}{9.1 \times 10^{-31} \times 7.7 \times 10^5}$$

$$\lambda = 9.4 \times 10^{-10}\,m$$